THE DARKNESS ECHOING

Also by Gillian O'Brien

BLOOD RUNS GREEN:
The Murder That Transfixed Gilded Age Chicago

THE
DARKNESS
ECHOING

EXPLORING IRELAND'S PLACES OF
FAMINE, DEATH AND REBELLION

Gillian O'Brien

PENGUIN BOOKS

TRANSWORLD PUBLISHERS
Penguin Random House, One Embassy Gardens,
8 Viaduct Gardens, London SW11 7BW
www.penguin.co.uk

Transworld is part of the Penguin Random House group of companies
whose addresses can be found at global.penguinrandomhouse.com

Penguin
Random House
UK

First published in the UK and Ireland in 2020 by Doubleday Ireland
an imprint of Transworld Publishers
Penguin paperback edition published 2023

Maps by Liane Payne

A CIP catalogue record for this book
is available from the British Library.

ISBN
9781529176957

Typeset in 10.8/15.75pt Minion Pro by Jouve (UK), Milton Keynes.
Printed and bound in Great Britain by Clays Ltd, Elcograf S.p.A.

The authorized representative in the EEA is Penguin Random House Ireland,
Morrison Chambers, 32 Nassau Street, Dublin D02 YH68.

Penguin Random House is committed to a sustainable future
for our business, our readers and our planet. This book is made
from Forest Stewardship Council® certified paper.

MIX
Paper from
responsible sources
FSC® C018179

For Alistair Daniel and in memory of Mai Crowe

CONTENTS

Welcome

'Any true book on Ireland must be
full of contradictions and thoughts
that may be only half true'
H. V. MORTON,
In Search of Ireland

My grandmother always said she'd make a handsome corpse. And when I saw her laid out on the bed, her hands neatly folded on her stomach, rosary beads entwined in her fingers, her soft leather shoes pointing to the ceiling, I knew she had been right. She was wearing the dress she'd had made for her son's wedding: an elaborate, full-skirted creation in russet, gold and black on which abstract geometric patterns competed with swirling floral motifs – it looked like a painting by Gustav Klimt. Her eyes were closed. The lamp on the bedside table threw an eerie glow across her pale face. I stood in the doorway, stunned. The clock on the bedside table marked time in languid ticks.

My grandmother opened one eye. 'Well?' she said. 'What do you think?'

Two nights earlier, over tea in front of the open fire, my grand-mother had outlined her plans for her funeral. This was not the first time – I'd been receiving regular updates about this funeral for years. I knew her choice of undertaker, how she wanted to be waked and what prayers she had chosen for her memorial card. On this occasion she wanted to discuss her 'rig-out'. Though far from rich, my grandmother had always been an extremely styl-ish woman, immaculately turned out in well-made clothes, and her funeral was to be no exception. Her plan, she told me, was to be laid out in the Klimt dress.

'But Nana,' I said, 'it's twenty years old.'

Nana did not see the problem.

'Besides,' I added, 'you're a much smaller woman now.'

Nana waved a hand. 'Yerra, it'll be fine. Shure Kirwan's will sort all that out.'

Kirwan's were the undertakers she had chosen. I was not con-vinced that their seamstress – assuming they had one – would be quite as nimble-fingered as Nana believed, and it seems my oppos-ition sowed the seeds of doubt. Two days later I arrived home from college to find her resplendently, if temporarily, 'dead'.

'Take a photo,' she demanded. 'I want to see how I look.'

Conversations with Nana about death were as ordinary as conversations about the weather. If you asked her how she was, she would invariably answer, 'Dying away,' a process that – though she was in perfect health – seemed to overshadow the last forty years of her life. And if it wasn't her own death, it was other people's. One day, when she returned from one of the innumerable funerals she attended, I asked her if her friend

Mrs Burke had been there too. She replied, 'Arra, she was, barely. That woman would be dead years if she'd the sense to stiffen.' On another occasion I took her on a trip back to her home town in Clare. I sat in a sugán chair by the fireside eating Swiss roll and drinking cups of tea so strong, as Nana used to say, you could trot a mouse across them, and listened as she and her old schoolfriend Catherine recited a litany of illnesses, deaths and funerals in the parish. As Catherine paused after describing the most recent funeral Nana sighed and quietly murmured, 'Shure, the whole world is dead.'

Nana was blithely unaware of her morbid nature. When I'd point out that she was preoccupied by death (hers and other people's) she'd dismiss my observation with a shake of her head and a 'Yerra, no more than anyone else.' And maybe that was the truth. As a nation, the Irish are obsessed with suffering and death. Where else would a website that publishes death notices get five million hits a month? Where else would a marriage proposal begin with 'Do you want to be buried with my people?' or popular ballads extol the attractions of martyrdom? From birth, the Irish are buried alive in a never-ending avalanche of stories, ballads, plays, wakes, dirges, poems, paintings, ceremonies, museums and heritage sites that celebrate death, misery and the macabre. From the tribulations of Peig Sayers to the plays of Samuel Beckett, we are reared on tales of desolation, oppression, failed rebellions, famine, civil wars and emigration, with the occasional sporting victory to keep our spirits up. Perhaps nowhere else on earth is *history* so inextricably entwined with *story* – a story of a very dark kind.

Stories about death have often been a source of levity and humour. Comic songs, plays and tales abound, from the ballad 'Finnegan's Wake' to Máirtín Ó Cadhain's *Cré na Cille* to Lisa McGee's *Derry Girls*. *Cré na Cille* is set in a graveyard where the dead gossip and feud and worry about how well they looked as a corpse and how impressive their funeral was. An episode of *Derry Girls* has a funeral as its centrepiece, complete with a scene about the etiquette of wake sandwiches. And obsession with death isn't confined to the Irish on the island. It's an obsession that travels well. My friend Jan grew up in an Irish family in Liverpool, and in the late fifties, like many others, the family regularly headed back across the Irish Sea to the Red Island holiday camp in Skerries (my home town). At the end of their holiday they took part in the fancy-dress competition, where they created an elaborate funeral tableau entitled 'Finnegan's Wake', which, as Jan described it, involved 'my big fat Grandad Billy as the corpse and my little dumpy Grandma Susie as the weeping widow'.

It took a move to Liverpool, marriage to an Englishman and attendance at some English funerals to make me fully aware that the Irish fascination with the dark side of a story was not shared across the globe. In England, death is unmentionable. I remember being horrified when Al, my husband, told me that when his mother died not one of his friends went to her funeral and when he returned to university everyone behaved as if nothing had happened. I'm still surprised by the formality of an English funeral, and after I'd been to a few I announced to Al that if I died in England his first action was to ensure that I was

4

repatriated to Ireland to be waked in traditional Irish style. What I didn't do was explain what that entailed. When my Uncle Seán died, Al and I went to his wake. We were welcomed at the door by my aunt and cousins and ushered into the sitting room, which was bustling with people. And there was my uncle, very much at the centre of things, just as he had been in life. This time, however, he was laid out in an open casket. I walked over and placed my hand on his cold one and gazed down. It's traditional at this point to observe how well a corpse looks, how they look 'just like themselves'. I turned to make some such comment to Al, only to discover that he was no longer behind me. He was propping himself up against the door lintel and looking more than a little green. 'You never said that Seán would be here,' he hissed when I crossed the room to retrieve him. It was true. It hadn't occurred to me to forewarn him that Seán's body would not be tidied away in a funeral parlour overnight, but be at home. After all, it was his wake. He wouldn't have wanted to miss the craic.

But while we Irish pride ourselves on providing a good send-off for the dead, we certainly don't have the most lavish death rituals – that accolade is probably best reserved for the Torajans, who live on the Indonesian island of Sulawesi. There, dead relatives are mummified and kept at home for weeks, months, sometimes years. The dead remain an integral part of the family, often chatted to, and provided with food and drink, their bodies washed and their clothes changed regularly. Eventually, following an extensive feast involving the ritual slaughter of a buffalo, the body is placed in an ancestral tomb. By contrast, an

Irish funeral might seem hasty, moribund and a little bit dull – though my uncle's was undoubtedly too exciting for Al.

How to explain the Irish obsession with death and misery? Is it the incessantly gloomy weather, or does it have something to do with 'national character' (whatever that is)? Undoubtedly, Mrs Doyle from *Father Ted* spoke for vast swathes of Irish people when, disgusted by the TeaMaster salesman who assured her that his contraption took the misery out of making tea, she retorted, 'Maybe I *like* the misery.' Heinrich Böll, the German writer and winner of the Nobel Prize for Literature, spent much of the fifties and sixties living on Achill Island off the coast of Mayo. In his *Irish Journal* he recounted a conversation with a man he met in a pub who asked him if he thought the Irish were a happy people: '"I think," I said, "that you are happier than you know. And if you knew how happy you are you would find a reason for being unhappy. You have many reasons for being unhappy, but you also love the poetry of unhappiness."' Maybe we do love the poetry of unhappiness, but I think it's more than that. The inclination to see shadows, not the sun, is rooted in a bleak past.

There's no escaping the fact that Irish history is full of darkness, and perhaps it's this history which encourages a fascination with the morbid, the melancholic, the miserable, the maudlin (so many 'M' words). These 'M's are the oxygen of conversation among the Irish. What else would we talk about? I'm sure I'm not the only one for whom every phone call home invariably features discussion about the weather, illnesses in the parish, the ridiculous price of houses, inept politicians and the appalling state of the health service. But if we took out all the melancholia, how would we navigate conversations with friends, family and strangers? There'd be long silences, and there's nothing the Irish fear more than silence. Misery provides our conversational staging posts, our pillars of chat. To some extent, we revel in the wretchedness, as Böll suggested; we delight in black humour and take pleasure in the laughter that often comes as a side order with grief.

The Irish do have a perverse attraction to nurturing their injuries. We don't need to have experienced terrible events, we just need to have heard about them to have absorbed some of the outrage from the past. The English travel writer H. V. Morton was near the mark when he visited Ireland in the late twenties and observed:

> The Irish are, of course, sometimes unfair, which I think proceeds from the fact that they possess no sense of historical perspective. Even educated Irishmen will talk about Cromwell's campaign as though it was the work of the present British Government. A wrong has never died in Ireland. Every injustice inflicted on Ireland since the time of Strongbow is as real as last year's budget.

And it's certainly true, as Professor Liam Kennedy has written, that 'the motif of victimhood bulks large in Irish national self-consciousness'. We've played the MOPE card many times; in fact, we have, if not a whole deck of cards, at least a hand, in which the Famine serves as our trump. And though we're not the Most Oppressed People Ever, much of the history of Ireland is indeed traumatic. Like the Sacred Heart hung over the mantelpiece, Irish history is an open wound, an object of veneration, proudly on display to all comers. And this is reflected in the stories we tell ourselves, not just when sitting in pubs and cafés, but also at times in our museums, at our heritage sites and in the pages of our history books. They are powerful tales with a strong narrative arc: first, suffering (a lot of suffering), then (hopefully) redemption.

Like many countries, Ireland has been anthropomorphized (depending on the political purpose) into several different versions of womanhood – in poetry, stories, art, sculpture and song, the country is portrayed as everything from an idealized young woman to a shrill old hag. 'Hibernia' is beautiful, noble and steadfast, often accompanied by her harp and her wolfhound. But throughout the nineteenth century she was often the 'injured lady', depicted in chains, an abused and traumatized woman (in many ways an accurate representation of the way real women were treated in Ireland, both before and after independence). And while it's sweeping to sum up a nation in the body of one fictional woman, there is something in the idea of Ireland as a traumatized nation. Colonization was a national trauma, and its effect on the body politic is not dissimilar to that on traumatized individuals. Clinical psychiatrist Judith Herman has argued that 'like traumatized individuals, traumatized countries need to remember, grieve, and atone for their wrongs in order to avoid reliving them'. And there is certainly a tendency in Ireland to continually pick at scabs. At the same time, repeatedly reopening old wounds leaves little space to consider the ones we might have inflicted on ourselves or others. But the way in which our history has been taught – formally and informally – for nearly two centuries has meant that the past, a past where the Irish are always victims, is almost always very present.

All national identities are a construct, fictionalized creations generated to serve some political end. And there isn't just one national identity – there are many, and they have many purposes, some pluralist and inclusive, others exclusive and divisive. In the 1840s, the Young Irelanders constructed an Irish national identity which evoked a heroic (if characteristically miserable) past. As James Quinn, author of *Young Ireland and the Writing of Irish History*, put it, they created a version of Irish history full of 'sufferings bravely endured, resistance that had never faltered and a national spirit that had never been crushed'. That remained the popular, and in many ways official, version until the late twentieth century, and there are those who still adhere to it today.

For years, Irish history, as taught in schools in the Free State and later the Republic, created a national history that was a nationalist history. In 1934, the Department of Education's 'Notes for Teachers' suggested that if history was 'properly' taught in primary schools, pupils would learn that they are members of 'a race that has survived a millennium of grievous struggle and persecution'. From the establishment of the Irish Free State in 1922 until the educational reforms in the mid-sixties, Irish history was used, as historian John O'Callaghan noted, for political objectives: 'the subject of history taught young learners a monolithic nationalist, anti-British and pro-Catholic history that was heavily dependent upon allegory and collective memory'. I recall learning Irish republican ballads and poems in school in the eighties – I can still recite sections of 'Renunciation', a homage to blood sacrifice written by Patrick Pearse, one of the leaders of the 1916 Easter Rising. In retrospect, it seems a little inappropriate to have had a class of ten-year-olds reciting, 'I have turned my face/ To this road before me,/ To the deed that I see/ And the death I shall die.'

The more modern (and Instagram-friendly) version of Irish history is that peddled by those organizations desperate to lure tourists to our shores. The tourism package has created an 'Ireland of the welcomes', a céad míle fáilte, a place of saints and scholars, of seanchaís and banshees. But broad brushstrokes paint an uncomplicated picture – both the political and the touristic version are partial and simplistic. They iron out all nuance and some narratives are excluded because there is a hierarchy around what is remembered and memorialized. Certain anniversaries, centenaries and bicentenaries are deemed more politically palatable than others and therefore become more visible.

As a historian, I've spent many years researching the murkier corners of Irish history. I've written a book about the gruesome murder of an Irish republican in nineteenth-century Chicago and I've worked as the historical consultant for a number of heritage projects – former prisons, courthouses and forts, places where stories of murder, conflict and incarceration abound. Working on these places stimulated my interest in what is called 'dark tourism' – travel to places associated with suffering or death. I wondered if our museums and heritage sites pander to these simplistic versions of Irish history and how they walk the fine line between catering to visitors' fascination with the macabre and providing the complex and sophisticated narrative required to explain the past without exploiting it. And so I decided to set out on my own bespoke tour of Ireland – a tour my grandmother would revel in (without admitting she was doing any such thing), a tour of Ireland in the shadows.

Although dark tourism as a concept is relatively new, the phenomenon itself is anything but. For centuries, people have been visiting sites associated with suffering, desolation, incarceration and death for their entertainment. In the seventeenth century anatomical theatres were as much public performances as they were anatomy lessons. In the eighteenth and nineteenth centuries thousands flocked to public executions. Madame Tussaud made a career out of grisly spectacles, opening her first Chamber of Horrors in London in 1802, while in the United States 'dime museums' – a mix of educational museum and freak show – attracted customers by advertising exhibitions where the paraphernalia of torture could be seen. Unsurprisingly, Ireland was not immune to the lure of the macabre; just behind the Four Courts in Dublin the mummified bodies in the vaults of St Michan's Church have been attracting visitors since the early nineteenth century.

There is a perception that sites which embrace the term 'dark tourism' both sensationalize and trivialize the stories being told and that they prioritize entertainment over education. In the public mind, the term is sometimes associated with voyeurism, tastelessness and vulgarity. But dark tourism can be much more nuanced than that. There are indeed tensions between entertainment and education, and complex decisions have to be made about what stories are told, how to tell them and what artefacts should be displayed but, even so, I think museum directors, curators and heritage-site managers should embrace the phenomenon. After all, they are responsible for tourist sites and these sites offer stories that involve misery, suffering, incarceration and death. That is dark tourism, like it or not.

Dark-tourism expert Philip Stone has identified many shades of darkness – the darkest being sites of unimaginable suffering and torment, like Auschwitz-Birkenau, where the focus is on education and historic interpretation, while the lightest are sites with a greater focus on (grisly) entertainment, such as the Amsterdam Dungeon, which promises to make visitors 'scream and laugh' their way through the city's darkest history. Irish sites cover most of the spectrum. The darkest of them is Kilmainham Gaol. As the prison which held many Irish nationalist heroes, from key figures associated with the 1798 Rebellion to the leaders of the 1916 Easter Rising, Kilmainham Gaol occupies a unique place in the Irish imagination, poised somewhere between a prison and a shrine. You won't find stag parties here (as I did on my visit to Crumlin Road Gaol in Belfast). At the lighter end of the scale is Leap Castle in County Offaly, where visitors are regaled with ghost stories. And for the well-researched dark tourist there is darkness to be found at sites that gloss over the grimmer aspects of their past. For example, at Desmond Hall in Newcastle West, County Limerick, the focus is firmly on banqueting and architectural features with not a mention of the soldiers who were executed there in 1643, their rotting bodies placed on stakes outside the building as a warning to others.

It's not easy being a dark tourist in Ireland. There's no 'Shadowlands' trail to sit alongside Fáilte Ireland's 'Wild Atlantic Way', 'Ireland's Ancient East' and 'Ireland's Hidden Heartlands'. Brand Ireland is a mythical land full of slogans and sun and bright, welcoming smiles. But I prefer the darker, damper corners. The real Ireland. So I bought a map and began to plot my own bespoke route around the country. To begin, I drew up a shortlist of sites of incarceration – prisons and workhouses. There were fewer than twenty places to see, or so I thought. It wouldn't take that long. A quick scoot to Wicklow and Kilmainham jails, a nip up the M1 to Crumlin Road Gaol, a couple of days in Cork at Spike Island and the city jail taking in a workhouse or two on my way. But I was easily lured off course and if there was another museum or gallery close by I popped in there too. Very quickly I realized that almost all the stories being told were about conflict, incarceration, famine, emigration or death. Even art galleries are stuffed with suffering – is there a gallery that doesn't have art depicting battles, rapes or crucifixions? Twenty sites rapidly became fifty, and then (thanks to suggestions from friends, family, the listeners of RTÉ Radio 1's *Ryan Tubridy Show* and people on Twitter) it expanded to well over one hundred and fifty.

A few months into what I began to call 'the Misery Project', I participated in a discussion at Dublin Castle about how the Famine has been represented in art. A woman in the audience asked if I thought visiting places associated with suffering or death was affecting me psychologically. Until then I hadn't given that much consideration, but I suggested that a combination of professional detachment and years living with my death-obsessed grandmother had immunized me against strong emotional reactions to tragic tales. Later, as my tour progressed, I often reflected on that question, for there certainly were times of sorrow on my travels, but also periods of joy and laughter.

The Darkness Echoing is much more than a guide to Irish museums and heritage sites (or to Irish history). It's a personal exploration of what stories are told and why. I'm particularly interested in the lesser-told tales, those that are overlooked and often sidelined while the same old stories take centre stage time and time again. But there are other angles, other voices to amplify, other artefacts to highlight, and with close inspection and a little digging around, gems are often revealed. In choosing where to go I've had to be selective. Site-specific attractions – the places where sad or bad things actually happened – have formed the nucleus of my journey (though I've visited plenty more traditional museums on my way). It has long been a USP of the site-specific museum that it offers something that the tourist just can't get anywhere else: immersion – a three-dimensional dunking in authentic. In our globalized age, in which everywhere resembles everywhere else, the unmediated authenticity of being in the place where significant events occurred exerts a powerful appeal. And so my travels took me to prisons, workhouses, castles and forts, but also to the landscape of loss – to battlefields and cemeteries – where thousands died and were buried. I followed the (hi)story of Ireland through its songs, its literature, its art, its buildings and burials. I've waded out to an island and climbed a Martello tower, been imprisoned in a jail and spent three nights in a former lunatic asylum, pausing only for breath and the regular sampling of scones.

When foreigners think of tragedy and trauma in Ireland they often think of the Troubles, but to understand our obsession with darkness it's necessary to go much further back, beyond the Civil War, beyond the 1916 Rising, beyond the Famine, beyond Cromwell's massacres, all the way to the stories of Cúchulainn and Queen Maeve. It's a combination of all these stories, real, exaggerated and sometimes entirely fabricated, that has made – and continues to make – us who we are. This book concentrates on events between the late sixteenth and early twentieth centuries, but slips either side of these at times. Megalithic tombs and graveyards, bog bodies and murals in Belfast and Derry, all make appearances, though I've been careful to avoid focussing on stories about people who may be still alive, or have relatives and friends still living.

I put a dot on my map to mark every site I planned to see. It began to look very cluttered indeed. After each journey I spread the map across the table, flattening it out carefully, making sure not to rip it. The creases became fragile from unfolding and refolding. Every evening I took out my red marker and traced that day's route along motorways, national, regional and local roads. It was my reward: another part of the country covered, another bit coloured in. As I stood back to admire my work I realized that the map had begun to resemble a diagram of veins, arteries of blood radiating from Skerries across the country. Far from tracing death, it looked as if I was slowly bringing the country to life.

I brought lots of people with me on my journey. For several years, Al has eschewed exotic foreign holidays in favour of traipsing after me at many historic sites. I also inveigled friends and relations to come with me on some trips but, alongside Al, my most regular travel companions were my nephew Jack, my nieces Abbey and Lucy, and Misha and Matilda, the children of my friend Jessie. Instead of taking trips to Taytopark or Disneyland Paris, they were persuaded to accompany me to prisons and cemeteries, to gaze at mummified bodies, to clamber around castles and forts and descend into caves. I was sure that my nephew and nieces were fit for the task. Abbey was already campaigning for me to take her to see Anne Frank's house in Amsterdam, while Lucy heard about the Dunbrody Famine Ship at school and demanded that I add it to the list. I was warier of taking Misha and Matilda along, but when Jessie told me that, as a young child, Misha had been entirely unfazed when he accidently unearthed some human remains while playing in his grandparents' garden (which happened to be an old churchyard), I was confident I'd picked the right kids for the job.

Despite a regular supply of willing (and unwilling) accomplices, for the most part I've travelled alone, though I've never quite felt alone, for my quest has been both inspired and haunted by a quartet of ancestors who I've come to think of as 'the four horsemen of the apocalypse' (although two of them are women). There's my grandfather, Denis, who insisted that the lyrics to

'Boolavogue' were a documentary account of the 1798 Rebellion; my great-grandfather, Jack, who fought in the 1916 Rising, and my great-grandmother, Bridget, who, when a robin flew into the house, declared it an omen of death and told us we might as well all give up living immediately. Their spirit(s) have infused much of this book.

The final member of the quartet is my death-obsessed grandmother, Mai. I never did take a photo of her laid out in her seventies finery. I had no camera with me that day, and when she did die, some fifteen years later, in her ninety-eighth year, she was dressed – to my relief – in the fine woollen suit and pink silk blouse she'd bought for her ninetieth birthday. Still, I don't need a photograph to remind me of that moment. I can see it in my mind's eye. But unless I put the story down on paper, or retell it to friends, it will eventually be lost. That, of course, is true of history itself. Some of it is written down, some of it survives through objects, some in bricks and mortar – but so much of it simply evaporates into thin air. Or, like ourselves, is buried or burned. What's left behind is all we know. In this book I set out to explore that fragile legacy, to ask what Ireland's dark history means to us, why we value these stories, and what they reveal about who we were, who we are, and who we could be. It's a personal odyssey into the dark heart of Ireland: its past, its present, and its psyche; into the stories we tell ourselves and what they mean.

Battles
and Sieges

❧

'Warfare of six hundred years!
Epochs mark'd with blood and tears!'
— WILLIAM DRENNAN,
Wake of William Orr

N

6 Siege Museum ■

Stormont ■

4

7 Oldbridge House ■ ■ St Peter's Church

9 Athlone Castle ■

1 National Gallery

Battle of Aughrim ■
Visitor Centre

2 Carrigafoyle Castle
■ King John's Castle 10
■ The Treaty Stone

■ Tipperary County Museum 5

■ Cahir Castle

Waterford Cathedral ■

■ Dun an Oir 3

8 Elizabeth Fort
■
St Fin Barre's Cathedral

If you're going to educate an Englishman in the history of Ireland, there's no better place to start than the National Gallery of Ireland on Dublin's Merrion Square. Specifically, the Shaw Room, in front of Daniel Maclise's iconic history painting *The Marriage of Strongbow and Aoife*. It's important to start here, I explained to Al, positioning him in front of the painting for the clearest possible view, because Maclise's painting symbolizes the beginning of the English conquest of Ireland, the start of those 'eight hundred years of oppression'. Over the course of our trips around the country Al would have to get used to hearing stories about how everything bad that ever happened to Ireland was England's fault. In the gallery I patiently outline the three main reasons why England gets blamed for all Ireland's ills:

1. It suits the Irish to believe that everything is England's fault.

2. The only foreign visitors Irish museums cater for are Americans, who tend to agree that everything was

England's fault (and are similarly untroubled by distinctions between England and Britain).

3. Everything really was England's fault.

At first glance *The Marriage of Strongbow and Aoife* seems like a classic example of nineteenth-century romanticism. At the centre stands Strongbow (Richard de Clare, Earl of Pembroke), clad in armour with a garland of laurel leaves around his head, taking the hand of Aoife, who, wearing a gold-embroidered cloak and tiara, looks demurely at the ground. But upon closer inspection that impression quickly fades. The bride's father, Dermot MacMurrough, King of Leinster, looks on with an expression of great scepticism while the Irish chieftains gathered behind him wear thick metal collars and chains around their necks. The wedding party is surrounded by a burning castle, the dead bodies of slain Irish warriors, lamenting women and, in one corner, an Irish musician slumped dejectedly over his harp. The symbolism can hardly be missed (except, perhaps, by the man who recently proposed to his girlfriend in front of the painting): for Ireland, the marriage of Strongbow and Aoife was a disaster that paved the way for centuries of colonial subjugation.

'Oh dear,' said Al.

Yet one of the most interesting things about the painting is that it was never meant to be viewed as a tragedy. As historian Tom Dunne has pointed out, Maclise was a unionist, and intended his painting to be displayed in the Palace of Westminster. To Maclise the painting was a celebration not of romantic union but of conquest (both territorial and sexual), a record of the violent origins of the British Empire. Surprisingly, the National Gallery, which was gifted it in 1879, didn't see it that way, and ever since the painting was included in *Cuimnechán 1916* – the gallery's exhibition to mark the fiftieth anniversary of the 1916 Easter Rising – it has been popularly regarded in Ireland as an unflinching exposé of colonial oppression filled with defiant symbols of Celtic Ireland.

The fateful marriage took place in the midst of internal power struggles in Ireland. In 1169 mercenaries sent by Strongbow arrived to assist Dermot MacMurrough in his quest to secure his Leinster crown and become High King of Ireland. The soldiers were followed a year later by Strongbow himself, keen to capitalize on one part of the deal – marriage to Aoife, MacMurrough's daughter. Within a year of the marriage MacMurrough was dead and Strongbow had become King of Leinster. The growing power of Strongbow provided the impetus for Henry II, King of England, to travel to Ireland to establish his authority there. By the mid-1170s, Henry II and his allies had taken control of large parts of the island (at least theoretically).

My friend Reachbha told me that at school she learned that Strongbow had enormously long arms. I was certain that she'd conflated the stories of Strongbow and Mr Tickle, but after some digging around she triumphantly unearthed an early-seventeenth-century source which noted that Strongbow 'had such long arms that with the palms of his hand he would touch his knees'. There was no sign of unusually long arms in Maclise's painting, nor on the effigy carved on Strongbow's tomb in Christ Church Cathedral in Dublin. However, despite a modern sign reading 'Strongbow' it turns out that the effigy is that of some other, unidentified, knight. Strongbow died in 1176 and was buried in the cathedral, but in 1562 part of the cathedral roof and walls collapsed and his tomb was destroyed. Later a substitute tomb took its place. But I did find one recent representation to bear out Reachbha's claims. Just outside Christ Church Cathedral in Waterford there are two bronze statues of Strongbow and Aoife commemorating the fact that their wedding took place in the city. Strongbow's arms are enormously long, though admittedly this is because they form the backrest of a chair. Strongbow looks utterly miserable while Aoife looks benignly resigned – having weary pedestrians come and sit on their laps is probably not how either of them thought they'd be remembered.

Reachbha's Strongbow story got me thinking. She was sure she'd heard it in school, and I wondered what other random 'facts' had lodged inexplicably in people's minds long after they'd last skipped out a school gate. History is (thankfully) a compulsory subject in Irish education until students complete the junior cycle, which means most people are formally taught Irish history between the ages of five and fifteen. And so I conducted an (entirely unscientific) survey, asking friends, family and pretty much anyone I could find what historical 'facts' they recalled from their schooldays. The most popular answers were:

Walter Raleigh brought the potato to Ireland.

The walls of Limerick could be destroyed by throwing rotten apples at them.

Wolfe Tone was murdered by an English soldier.

You had to be Catholic to be properly Irish.

All landlords were English.

Queen Victoria gave £5 in famine aid.

De Valera ordered Michael Collins's assassination.

None of these 'facts' are true, and I was curious to discover to what extent these myths persist in Ireland today.

From the late twelfth century onwards, the English Crown attempted to consolidate its control of parts of Ireland and was met by fierce, if intermittent, resistance. The island was a landscape of constantly shifting alliances primarily driven by tribal concerns. Vicious internecine struggles between Irish chieftains dominated. Interestingly, most Irish museums tend to gloss over these. In Ireland we prefer rebellion over civil war. In the late sixteenth century the country was still a jigsaw of complex and ill-fitting affiliations, where loyalties ebbed and flowed on a regular basis. Studying this period as an undergraduate was something of a trial. Just as I'd begin to understand who was fighting who and why, the deck of cards would reshuffle itself and I'd have to start all over again.

I wrote part of this book on the Dingle Peninsula, and while in Kerry I decided to visit some sites associated with the Desmond Rebellion of the early 1580s. In the late sixteenth century, over a hundred years before William and James were slugging it out for the Crown, and sixty years before Cromwell wreaked havoc across the country, Desmond's Rebellion made an early but plausible bid for Ireland's Most Confusing Conflict, a heady mix of religious war, power struggle and anti-colonial uprising.

Under Queen Elizabeth I, England had continued to spread its power and influence beyond the Pale, attempting to subdue the entire country. Munster – as anyone from there will proudly tell you – was the province that put up the stiffest resistance. Gerald FitzGerald, Earl of Desmond, and his cousin James Fitz-maurice FitzGerald rose against the Elizabethan forces, appealing to Catholic Europe for support: a strategy that was to prove strangely popular over the centuries, despite a string of less than impressive results. In fact, Desmond managed to attract only a few hundred Spanish and Italian mercenaries to his cause. Against the might of the Crown, Desmond's rag-tag army stood little chance.

The remains of this brutal period are scattered across the south-west of the country. One site where the rebellion sprang bloodily to life was at Carrigafoyle Castle near Ballylongford in north Kerry. When it was built in the late fifteenth century the castle was one of the strongest fortresses in the country, exerting considerable control over the shipping making its way up and down the River Shannon to Limerick. The fortified walls, part ruined by time and a besieging army, are surrounded by a reedy estuary which formed a natural moat. The castle itself is a single, five-storey rectangular tower, through which English cannon blew a gaping hole at Easter 1580, creating the sort of cross-section you find in a Dorling Kindersley book illustration designed to show the inner workings of a castle.

I regretted the fact that I hadn't visited sooner, as the previous week I'd spent an evening with my ten-year-old niece, Lucy. Lucy has two passions – history and gymnastics – and when she wasn't showing me a backflip, a cartwheel or some other gymnastic feat, she was constructing her own castle for a school project. Made of cardboard tubes and empty cereal packets, Lucy's castle was encircled by a fence made of ice-pop sticks and it was an imposing sight on the kitchen countertop. My contributions had been limited to 'maybe you should paint it grey' after she'd already committed to brown, and 'add a few more flags and banners'. Now, standing at the entrance to Carrigafoyle Castle, I realized that I should have been encouraging her to add an impressive moat, or more fortifications to her walls, while wrestling with double-sided sticky-tape and a paintbrush.

Al and I made our way into the tower, now an unofficial bird sanctuary inhabited by resentful swallows, and followed a handful of jocular American tourists up the winding stone staircase to the battlements, where there is a commanding, and very scenic, view of the Shannon Estuary. Here, on the wind-whipped battlements in summer sun, I got a sense of what it must have been like to face overwhelming odds. Fifty yards away, just beyond the car park, I could clearly make out the grassy knoll where the English forces led by Sir William Pelham set their cannon – a range best described as point blank. I traced the cannonballs' trajectory through the huge hole blown in the outer wall, and the missing side of the tower below me suddenly made new and terrifying sense. The bombardment lasted three days before the castle surrendered. The garrison was executed and the Earl's valuables were sent to Queen Elizabeth.

The attack on Carrigafoyle was just one incident in a conflict that rumbled on for several years. In November 1580 several hundred Spanish and Italian mercenaries who had answered the enticing clink of the Earl of Desmond's gold coins found themselves surrounded by Elizabethan ground and naval forces on the Dingle Peninsula. In vain they attempted to build defences at Dún an Óir, near Smerwick harbour – the remains of hastily constructed ramparts and part-built bastions can be seen at the site of the original Iron-Age fort today. As at Carriga-foyle, their resistance lasted only three days before the beleaguered troops surrendered – more in hope than expectation – and were promptly massacred. Some sources claim that one of the English soldiers involved in the massacre was a young captain called Walter Raleigh. There were reports that many were tortured, their arms and legs broken in three places; hundreds were hanged, others beheaded.

Near the edge of the cliff overlooking the harbour these macabre events are commemorated by a somewhat literal sculpture of twelve disembodied heads, but there is very little information at the site so I visited Músaem Chorca Dhuibhne in Ballyferriter in the hope of filling in the gaps. In fact, the museum has only one small panel on the rebellion and I learned more from the museum's questionnaire for children. It is not a questionnaire for the faint-hearted. 'There are two fields near here called Gort a' Ghearradh (the field of the cutting) and Gort na gCeann (the field of the heads). Can you think why?' It was quite refreshing to find a worksheet for children that was more Horrible Histories than Ladybird guide.

For the most part, the Desmond Rebellion was haphazard and poorly planned, but the devastation across Munster was widespread, for both sides adopted a scorched-earth policy, destroying crops and livestock and resulting in the deaths of thousands of civilians from starvation and disease. The *Annals of the Four Masters* (a history of Ireland up to 1616) noted that 'it was commonly said that the lowing of a cow or the whistle of the ploughboy would scarcely be heard from Dunquin to Cashel'. The poet Edmund Spenser, who was secretary to Arthur Grey, Lord Deputy of Ireland, wrote that 'out of every corner of the wood and glens they came creeping forth on their hands . . . Anatomies [of] death, they spoke like ghosts crying out of their graves.' The Desmond Rebellion came to a bloody and bitter end when in late 1583 the Earl himself was captured near Tralee, not by Crown forces but by a local clan, the Moriartys – as so often in Irish history, it was the shifting sands of local partnerships that proved fatal. Desmond was decapitated and his head dispatched to Queen Elizabeth, who was presumably less keen on that parcel than she had been on the valuables sent from Carrigafoyle Castle.

The Desmond Rebellion may be complex, but for me no decade is more confusing than the 1640s, when waves of violence rolled across the country and it's estimated that between 20 and 40 per cent of the population died through warfare, famine or plague. It's another decade of tenuous alliances – the decade of the 1641 Rebellion, when Gaelic Ireland rebelled against the Plantation of

Ulster, which had forced thousands of Gaelic Irish off their land and replaced them with English and Scottish settlers who were loyal to the Crown. It's the decade of the Confederation of Kilkenny, when Irish Catholics tried to wrest back political and military power. And it's the decade of Oliver Cromwell. The drama of the 1640s is mentioned at sites across the country, but all I'm left with are snapshots and snippets of knowledge. But it's also confusing because much of the decade, as historian Tom Bartlett has observed, defies coherent description and is too complex to succinctly explain in a hundred words of panel text. Sites, quite sensibly, tend to focus on their local story.

Amid all the complexity, one thing I was certain of was that I wouldn't find any site or museum with a good word to say about Oliver Cromwell. In Britain, Cromwell is a much more ambiguous figure, lauded in some quarters as the father of republicanism (not a garland he would be given in Ireland). A statue of him, sculpted by Hamo Thornycroft in 1889, stands outside the House of Commons in Westminster. But his celebrity there has nothing to do with his exploits in Ireland, which are largely unknown (a general rule in Britain is that if it happened in Ireland, it doesn't count). In Irish history, however, Cromwell is always the villain. In August 1649, having won the Civil War and executed King Charles I, Cromwell and his New Model Army arrived in Ireland determined to take control of the country in the name of Parliament and to confiscate land from Catholics and royalists who had supported the King. This is not a conflict that can easily be wrapped in a green flag (though, as we'll see, many have tried). Over the course of the next few years Cromwell and his men wreaked havoc throughout the country, but the one event that Irish people know about, if they know anything at all about Cromwell in Ireland, is the siege of Drogheda, which took place in September 1649, a few weeks after Cromwell landed at Dublin. Drogheda was one of the best-fortified towns in the country, but it could not withstand the sustained assault. On 11 September, after more than a week of bombardment, the walls were breached by the Cromwellian forces and no quarter was given to those inside the town. It's estimated that several thousand were brutally killed – many of them were royalist soldiers, but hundreds of civilians also died.

As a child, I was taken on a school tour to Drogheda to see 'Oliver's head'. Retrospectively, this seems a bizarre thing to have done: pile a bunch of under-twelves on to a coach and drive them thirty kilometres to St Peter's Church to see the head of a man who had been decapitated. All I really knew about Drogheda was that at some point in the dim and distant past 'an Oliver' had massacred men, women and children in the town. Since no one ever mentioned a second Oliver, I had no idea that the head I was looking at was not Cromwell's but that of St Oliver Plunkett, the Catholic Archbishop of Armagh, who was executed in London in 1681 for plotting to assassinate King Charles II and reinstate Catholicism as the primary religion. Despite the fact that there was no substance to the charges, Plunkett was found guilty of treason and was hanged, drawn and quartered, and in death the Catholic Church – for reasons best known to itself – dismembered him a second time, dispersing his body in churches, monasteries and convents across Europe.

I know more now, so I decided to retrace my childhood steps and find out how Drogheda remembers the two Olivers today. In that slump between Christmas and New Year my friend Stephen took me on a tour of his home town. How better to fill the festive season than with a misery tour? We started at the old jail on Scarlet Street (so-called because of the blood that flowed down it following the Cromwellian attack), taking in the sites of execution (outside the Tholsel), then visited St Peter's Church to see St Oliver Plunkett's head. When I visited on my school tour, Plunkett's head was kept in a dark corner of the church in what looked like a tabernacle placed on a small altar. The front of the case could be opened and inside, behind a sheet of glass, visitors and pilgrims could catch a glimpse of a wizened face. Now, Plunkett's mummified head is encased in a lavish gold- and jewel-encrusted shrine with glass panels allowing visitors a very clear view of Plunkett's head from all angles. His eyelids are closed, but his lips are pulled back in what looks like a grimace and his teeth are clearly visible – it's quite a grotesque sight. The church also has his hip bone on display and, if that isn't enough to satisfy you, his thigh bone can be seen in the church in Oldcastle, County Meath, where he was born.

From the church we crossed the Boyne via St Mary's Bridge and walked up the hill to the 'Cup and Saucer' – the Martello tower which houses part of the Millmount Museum. Despite the fact that the town is most famous for its association with Oliver Cromwell, the museum makes little of the siege – there are a couple of panels in a stairwell and in the Martello tower itself the siege is dispensed with in three sentences. Outside the museum, the gaudy Ireland's Ancient East panel told us that Sir Arthur Aston, 'the unpopular one-legged defender of Drogheda', had his brains 'beaten out' with his own wooden leg by Cromwellian soldiers who believed his leg was full of gold coins – they were disappointed.

In 2000 a death mask of Oliver Cromwell was exhibited at the Drogheda Heritage Centre, but the exhibition was not without controversy. Protestors carried placards with slogans such as 'Cromwell drank Drogheda Blood' and 'Satan and Cromwell, lads together' (which made it sound as if they were part of a diabolical stag do on a pub crawl along West Street). Frank Godfrey, a town councillor and deputy mayor at the time, brandished a sword at the protest and claimed the death mask was 'a symbol of death and evil' (unlike the sword, apparently). In low-budget Hammer Horror style, tomato ketchup was smeared on the walls of the heritage centre to represent the blood of those killed by Cromwell's troops. Yet, despite, or perhaps because of, the protest, visitors flocked to see the death mask. Maybe the town should mention him more.

Cromwell's death mask is displayed without fanfare (or protest) in Tipperary County Museum in Clonmel. Following the siege of Drogheda, Cromwell and his army continued their military campaign across Ireland and, in the spring of 1650, an army of over 14,000 Cromwellian soldiers encircled Clonmel, where Hugh Dubh O'Neill commanded a force of 1,600 men. For weeks, Cromwell's troops subjected the town to relentless bombardment, but the walls held firm, and Cromwell was close to abandoning the siege when one of his soldiers brought him a silver bullet that had been fired from the town. Cromwell knew that if bullets were being made from silver, then the defenders had run out of lead and their ammunition would soon be exhausted. Encouraged, he resumed the bombardment and on 16 May 1650 his troops broke through the walls and ransacked the town.

Cromwell's brutality was as much strategic as bloodthirsty – he claimed that the savagery evident at Drogheda served as a warning, to 'prevent the effusion of blood for the future'. News of his army's atrocities did not take long to spread across the country, and many of his targets decided not to put up a fight. Cromwell wrote a short note to Lord Cahir demanding he surrender his castle in County Tipperary. It was not what you'd call a veiled threat: 'having brought the army and my cannon near this place, according to my usual manner in summoning places, I thought it right to offer you terms honourable to soldiers ... But if I be ... necessitated to bend my cannon upon you, you must expect the extremity usual in such cases. To avoid blood, this is offered to you by Your servant, Oliver Cromwell.' Lord Cahir duly surrendered, and both he and his forces were spared. Cahir got his castle back in 1662, following the restoration of the monarchy, and held on to it until his death in 1677 from 'a surfeit of claret'.

After Cromwell's death in 1658, the Commonwealth of England, Scotland and Ireland fractured and collapsed and in May 1660 the Stuarts returned to power in the shape of Charles II, son of the executed Charles I. When Charles II died in 1685, he was succeeded by his brother James. Within four years the tensions surrounding a Catholic monarch leading a Protestant state had boiled over in war. King James II was the first Catholic monarch since Queen Mary in the 1550s. His daughter Mary, a Protestant, had been next in line, but in June 1688 she was supplanted by the arrival of her half-brother (another James – royalty really need to be more imaginative with names, it causes havoc for historians), who James decided to raise as a Catholic. This was a horrific prospect for a political establishment that had spent most of the previous century embroiled in murderous wars of religion. What had been regarded as a brief interlude of Catholic power now began to assume dynastic potential. The Protestant elite, Mary, her husband William of the Dutch House of Orange, and other European leaders feared a change in the European balance of power. The 1689 coup in which William and Mary overthrew James and installed themselves as dual monarchs came to be known (in Britain at least) as the 'Glorious Revolution'. But James II did not go down without a fight and, as so often happened, Ireland became the battleground. There, with a largely Catholic population and senior Catholic figures in political and military positions, James was assured of considerable support.

The 'War of the Two Kings' or the 'Jacobite/Williamite War' (1689–91) is a war of many 'truths'. For unionists, it's their foundation myth. For nationalists, it's another black mark against Protestant English domination (though William wasn't English and James was). In Ireland, the war is seen as an Irish conflict, a Catholic-versus-Protestant conflict, a clear case of good versus bad (with each side claiming to be the good guys). But the reality is that it wasn't even about Ireland, it was a war about power, about who got to be King (or Queen – Mary is almost always written out of the story) of England, Ireland and Scotland, and if James had won he'd simply have begun another war to reclaim his English crown.

Alongside Catholic Irish support, James II also received assistance from King Louis XIV of France, which bolstered his army of Catholic Irish and Anglo-Irishmen, while the army of William of Orange was even more international, with Dutch, English, Scotch, Welsh, German, Danish, French Huguenot and Protestant Irish soldiers. Communicating battle plans to this diverse group must have been quite the challenge. The conflict took place in several locations throughout Ireland and the Grand Orange Lodge has produced 'The Williamite Trail' which identifies twenty key sites and a further thirty-one associated sites scattered across the island. I focussed on just a handful. The bullet points – or musket shots – if you will.

Tucked down a side street within Derry's (or Londonderry's) city walls is the Siege Museum. The museum is lodged in a newly built extension to the Apprentice Boys Memorial Hall, and if the man behind the reception desk was surprised to find a woman with a Dublin accent purchasing a ticket, he had the grace to conceal it. Al and I were the only visitors and, while I was distracted by some impressive trench art in a temporary exhibition on the First World War, Al wandered through the exhibits, admiring the 'no surrender' crockery, some of which had been overpainted recently in brighter, more orangey orange, and chuckling to himself as the loudspeakers played a reedy version of 'Can't Help Falling in Love with You' on the pipes.

The Siege of Derry is the most famous siege in a war that consisted almost entirely of them. On 7 December 1688 a group of apprentice boys raised the drawbridge and locked the city gates, keeping the army of James II outside the walls. Unlike in traditional sieges, James's supporters, the Jacobites, made no real attempt to storm the city walls; instead, they barricaded the River Foyle in order to cut off the city's supplies. Conditions within the walls deteriorated rapidly, and while there were some casualties from combat, the vast majority of those who perished in the city (up to ten thousand) died from starvation or disease. In the Siege Museum we found plastic rats scattered throughout the display; one erstwhile apprentice boy sported one jauntily on his head. During the siege, rats proved to be a popular delicacy, sold for a shilling apiece. A quarter of a dog cost five shillings and sixpence. Before the siege, sixpence would have bought 'a fat turkey', but during it all it would buy you was a mouse.

In a desperate attempt to alleviate the suffering and bring the siege to an end, Derry's Governor, Lieutenant-Colonel Robert Lundy, sought and secured support from King William. In return for his aid, everyone in military or civil positions had to swear an oath of allegiance to King William and Queen Mary. Lundy swore his oath privately, but refused to repeat it publicly, which led some to suspect disloyalty. By April he believed that defeat was inevitable and suggested the city surrender. There was little backing for this and on the night of 20 April 1689, disguised as a soldier, Lundy fled the city. Derry did not surrender and at the end of July, the Jacobite barricades across the Foyle were breached and Williamite ships carrying desperately needed supplies sailed on to relieve the city. Ever since, Lundy has been regarded as a traitor by unionists and loyalists and his effigy is burned on the first Saturday in December to mark the shutting of the gates and as a warning to traitors. The museum takes no great pains to balance its point of view (indeed it seems, in some respects, to serve as a recruitment centre for the Apprentice Boys), and the figure of Lundy on display is enticingly flammable. Visitors can also have their photo taken as 'Lundy the Traitor' by putting their head through a cut-out of him, rather reminiscent of the stock and pillories that litter prison museums.

I'm curious about this widespread belief that Lundy betrayed Derry. His motives may have been honourable. Believing there was no chance of victory, it's possible that he prioritized the lives of those in his beleaguered city. What we do know is that when he fled he did not join the Jacobites, he was never found guilty of treason and he continued to serve as a soldier in the Williamite army. This suggests, at least, that nothing is as simple as it might appear at first glance – but there's little room for nuance at a bonfire.

By the summer of 1690, both kings were in Ireland. On 12 July King William and his army of thirty-five thousand men took on the twenty-four thousand men loyal to James II at the Battle of the Boyne, undoubtedly the most famous battle in Irish history. What makes the battle most interesting is not the final score (James II, William III) but the fact that both kings were there.

Even in the seventeenth century, it was very unusual for monarchs to be anywhere near the action. And it's the iconic depiction of 'King Billy' crossing the Boyne astride his white charger that is imprinted in the popular consciousness, though it's unlikely that he was either on a white horse or forded the river.

The image of King Billy on a rearing horse has adorned both the gable ends of walls in unionist parts of Belfast and Derry and the banners used by the Orange Order. Indeed, it's even at the heart of Stormont – the Northern Irish Parliament building – where a painting depicting a heroic figure in full armour astride a white horse can be admired by visitors. The painting, by the Dutch artist Pieter van der Meulen, was bought, sight unseen, in 1933 and was intended to represent the firm unionist credentials of the Northern Irish Parliament. Those purchasing the painting believed it showed William III in Ireland, preparing for the Battle of the Boyne, and indeed the central figure is surrounded by well-wishers wearing orange sashes urging him onward to victory. But perched on a cloud in the sky above King William, and seemingly giving his blessing to the Protestant king's fight against Catholic King James, is the Pope. Within weeks of the painting's arrival in Stormont it had been vandalized by outraged visitors from the Scottish Protestant League, who threw red paint over the depiction of the Pope and slashed the canvas with a knife. Recent investigation by the art historian Bendor Grosvenor has discovered that this painting has nothing to do with King William or Ireland but is of St George being encouraged by a guild of Catholic crossbowmen in Antwerp as he set off to slay the dragon. While in Dublin *The Marriage of Strongbow and Aoife*, a piece of British imperialist triumphalism, has been reappropriated by some as a parable about the evils of empire, and by others as a romantic wedding scene, in Belfast a piece of Catholic mythology is presented as a celebration of Protestant military victory, which perhaps goes to show that the interpretation of art, as of history, is never final or fixed.

The story of the Battle of the Boyne is told at the site of the conflict. Oldbridge House, a few miles outside Drogheda, was the family home of the Coddingtons, who had been granted the land by William III as a reward for their support. The house, built in the 1750s, now hosts an exhibition about the battle. Jack, Abbey, Lucy and I took a trip to see it a few days after Christmas, and we were the only visitors to the house, though the surrounding battlefields were once again full, this time with joggers working off their Christmas excess. While there I recalled all the hullaballoo that surrounded Ian Paisley's visit to the site in May 2007, just days after his election as First Minister of Northern Ireland. Paisley visited alongside Bertie Ahern, then Taoiseach. While Ahern was empty-handed, Paisley came armed. Another showdown between ideologically opposed leaders, but this time the weapon was brandished peacefully. Paisley carried a musket used by the Jacobite side during the war, now (in the spirit of the time) a decommissioned weapon, a gesture linking a turbulent past to a peaceful present. The gift, which was accompanied by much fanfare, was proudly displayed in the museum until it turned out it wasn't a gift, but a loan. Eight years after the presentation the musket was returned to its owner (not Paisley, but someone in Antrim), who sold it at auction for £20,000.

Bolstered by mugs of hot chocolate from the coffee shop, we set off to explore the battlefield. There's not much to see – a few wooden-frame houses to indicate where the village of Oldbridge once stood, but little else. The children and I squelched our way across the grass until we found a little shelter underneath the spreading branches of a majestic five-hundred-year-old oak tree, a silent witness to the events of July 1690. Leaning against the trunk of the tree on that quiet December day with a soft rain falling through the branches, we looked across the huge green field flanked by trees and hedges and imagined the cacophony of battle, muskets firing, cannon roaring and blood spilling across the fields. While Lucy did cartwheels across the grass I explained that, most of the time, the soldiers would have found it almost impossible to identify who was on their side, as both armies wore red coats and blue trousers (which seems like an amateur mistake; surely they should have home and away kits – though both would probably have claimed to be the 'home' side). The only way to distinguish between the two armies was by the colour of the feather in their caps or a piece of ribbon or paper pinned to their sleeves. In the heat of battle this would have been a considerable challenge. Jack pondered the possibility of turning the battle into a computer game while Abbey ran her fingers around the trunk of the tree, wondering if there were any musket balls embedded in the bark – if there were, they had long ago been covered over by fresh growth.

In the wake of the Williamite victory at the Boyne, the Duke of Marlborough was dispatched by King William to secure Cork. One of Marlborough's targets was Elizabeth Fort, which is perched on a hill overlooking the south side of Cork city. When it was built in the early seventeenth century it afforded protection to the nearby South Gate, which controlled access to the walled city. The entrance to the fort is surprisingly well hidden amidst a warren of ancient streets that snakes up from the river. To its right is Barrack Street, then filled with inns and boarding houses and stables to accommodate and entertain visitors to the city. To the left is St Fin Barre's Cathedral. When Marlborough's men marched on Cork they positioned themselves on the surrounding hills. While in Derry it was the Jacobite soldiers that besieged the city, this time it was the turn of the Williamite forces. Outside the city walls some streets, including Barrack Street, were set on fire. Mortar shells rained down on the city and the fort. Within days the Jacobite soldiers were low on ammunition and, following a huge barrage of cannon fire, part of one of the fort's bastions collapsed and the city walls were breached. On 28 September, Cork surrendered, though that did not save the suburbs outside the city walls from being burned nor many of the Jacobite leaders from being summarily executed.

Evidence of the Williamite attack on Cork is hidden in plain sight across the city. After my tour of the fort I was in need of a short respite from battles and sieges and thought I'd sample the serenity of the neighbouring cathedral. I ambled through it, admiring the beautifully decorated chancel ceiling and being somewhat perplexed by the plaque to Elizabeth Aldworth, who was a member of the Freemasons (that notoriously male-only organization) during the eighteenth century. There was an organist rehearsing 'Lift

Up Your Heads' from Handel's *Messiah* and as I listened my gaze (perhaps prompted by the music) wandered upwards. Above my head, suspended from a chain, was a cannonball. It had been fired from Elizabeth Fort during the siege and had embedded itself in the steeple. When the steeple was demolished in 1864 it was discovered and, for some unfathomable reason, it now dangles from the ceiling, like a circular sword of Damocles.

After the Battle of the Boyne James licked his (metaphorical) wounds and returned to France (as did many of the French troops), but he did not give up. While Cork capitulated to the Williamites within days, Athlone, then a town of about 1,500 people, put up a more robust defence.

My visit to Athlone was a return to my childhood. My father is from there and many Sundays were spent driving for what felt like hours to see my great-grandmother in Coosan. The car trips were enlivened by the prospect of Gran bringing either a tin of boiled sweets (lightly dusted with sugar) or a big bag of Murray Mints. When we got to the house we would find Great-Granny Gorman, wearing her usual brown beret, sitting in a wooden armchair beside the fire. There was a fire burning in the grate (whatever the weather) and on one wall there was a Sacred Heart, illuminated by a small red bulb, on another a large framed copy of Walter Paget's painting of James Connolly lying injured in the General Post Office (GPO) during the 1916 Rising. It's no wonder so many Irish people are at ease with suffering, for their walls were lined with martyrs – secular and religious – showing off gory wounds. Tucked into the frame of the GPO print were paper Easter lilies, one more each year. Granny Gorman had tales aplenty about 'the Tans' and how they'd burned out houses and destroyed farms in Coosan in 1921 and how the Irish Republican Army (IRA) had destroyed Moydrum Castle, home of Lord Castlemaine, in retaliation. These stories were ancient history to me, but they were part of her lived experience. She hadn't told me any tales of earlier times, so I knew little else about the history of Athlone. It was time to remedy that.

I took Jessie's son, Misha, with me and, because there's nothing a sixteen-year-old boy would enjoy more than staying overnight in a convent, I arranged for us to stay with the Disciples of the Divine Master, whose convent overlooks the castle. Misha is fascinated by criminal psychology and forensics so to make it up to him we spent the evening at a talk about crime scenes and blood-spatter analysis, where he was put to work acting as a murder victim while I chalked his outline on the floor. The next day we walked down Church Street, crossing the River Shannon at the bridge. There has been a fortification on the castle site since the thirteenth century, but most of what remains today dates from around 1800, long after the siege. From the vantage point of the bridge the castle looks quite squat – like a small Martello tower. However, peering over the bridge, we could see that the defensive walls drop right down to the river, where it looks much more impenetrable, as it had to be, for Athlone Castle was the key fortress on the Shannon.

The museum inside the castle tells the history of Athlone and the castle itself, but the main focus is on two sieges that took place in 1690 and 1691, when the town briefly held the balance of power in Ireland. We made our way through the exhibition, where a series of animated maps projected on to the floor showed the progress of the two armies across Ireland, the Jacobites retreating from one town to another as William's forces closed in, and past the battle scenes re-enacted on screens all around us, until we reached the display on the siege. Here we found an interactive computer game where Misha and I were invited to play the besiegers. Having studied the Williamite wars in preparation for this book, I was confident that we would do well. We were asked how we would gain access to the castle and given three options: a. scale with ladders, b. pretend to be nuns, c. use the cannons. Given that we were staying in a convent, we naturally opted for 'b' and immediately caused the death of a hundred of our soldiers. In our defence, the Williamite forces found the going almost as tough in real life. The Jacobites, led by Colonel Richard Grace, successfully withstood the first siege, which took place in July 1690, just after the Battle of the Boyne. But in June the following year a much larger army led by General Godard van Reede van Ginkel began a second siege of the town.

I've always associated Athlone with the military, at least in part because my great-grandfather worked in Custume Barracks near the centre of the town. As a child, I thought it was called 'Costume Barracks' because soldiers' uniforms were costumes of a sort. It was only on my trip to the castle that I discovered the barracks were named after Sergeant Custume, who had fought on the side of King James. Aware that the Williamite forces were making efforts to repair the bridge across the Shannon (which the Jacobites had destroyed to protect themselves), Custume and ten other men from his dragoon regiment ventured out to sabotage the repairs and stop the Williamite advance.

There's a striking print on display in the castle which shows Custume and his men dressed in patriotic green. In reality, they would have been wearing red coats and blue trousers, but by 1910, when the print was made, recasting the Jacobite forces as unlikely nationalists was common practice. It's this patriotic version of Custume that entered the popular consciousness and may well have been responsible for the barracks in Athlone being named after him in 1922. In the event, Custume and his men were fatally exposed on the bridge and easily picked off by Van Ginkel's troops. Over the course of several days the Williamite soldiers used forty cannon to fire twelve thousand cannonballs, reducing the town to rubble and seriously damaging the castle. On 30 June 1691 Athlone fell to the Williamites, and the Jacobite troops, commanded by the French general Charles Chalmont, Marquis de Saint-Ruth, withdrew and headed further west. Misha and I followed them to Aughrim in County Galway, where, in a small one-storey building that looks like a repurposed community hall, we discovered that the Battle of Aughrim was a fleeting, bloody affair which began on the afternoon of 12 July 1691 and was over by the evening. Saint-Ruth was decapitated by a cannonball and much of the infantry massacred. The war was effectively over, but one final act had yet to be played out in Limerick.

I was the first visitor to King John's Castle on a bright but freezing December morning. During the war Limerick had remained loyal to King James and at the end of August 1690 Williamite troops attacked the castle. George Story, the Williamite chaplain, recalled that 'the noise was so terrible that one would have thought the very skies ready to rend in sunder [sic]. This was seconded with dust, smoke and all the terrors that the art of man could invent, to ruin and undo one another.' The Jacobite general, Comte de Lauzun, feared the castle walls were so flimsy they could be 'battered down by rotten apples', but it turned out that he was wrong: the walls held and the Williamites were repelled by a combination of Jacobite soldiers and local women, who hurled stones and bottles from the walls. But the supporters of King William were nothing if not persistent, as we saw in Athlone. A year later, soon after the Battle of Aughrim, they attacked again. The city was surrounded and bombarded with mortars, which tore huge gaps in the walls. On 3 October Patrick Sarsfield, Jacobite commander in Limerick, surrendered and the Treaty of Limerick was signed, bringing the war to an end.

The museum element of King John's Castle is in an adjoining building, and to enter the castle itself you walk through the undercroft before emerging into the courtyard. There I looked down at the entrance to a tunnel dug during one of the many sieges. Rather than the usual tactics of starving those inside the castle into submission, the besiegers decided to dig tunnels under it then destroy them, causing sections of the castle to collapse and allowing the attackers to gain entry. My experience of walking through the undercroft was a rather surreal one, for the entire space was festooned with Christmas lights and in the entrance to the siege tunnel were two large polar bears. A whole winter wonderland just for me, though it made the immersion in the grim history around me rather difficult.

My final stop in Limerick wasn't a museum but a large lump of limestone which sits on top of a fairly unprepossessing pedestal beside Thomond Bridge. It's the Treaty Stone, upon which the terms of the Treaty of Limerick were signed. These terms were generous to the vanquished – soldiers in James's army who had fought to the end were given several options. They could return home, they could join William's army, or they could go to Europe and join James's Irish Brigade or other continental armies. A clause was inserted giving Catholics the right to practise their faith and affording them protection 'from any disturbance upon account of their said religion'. But six years later, when the treaty was ratified by the Irish Parliament, this clause had disappeared and a frightened, insecure and vindictive Protestant parliament set a course that over the next three decades enshrined inequality in the country and created a situation where Presbyterians became second-class and Catholics third-class citizens (or, more correctly, subjects). These laws became known as the Penal Laws, and the implementation of some of them (and the simple existence of others) had serious repercussions for many Catholics and Presbyterians for generations.

Rebellion and Revolution

❧

'All changed, changed utterly:
A terrible beauty is born'
— W. B. YEATS,
Easter 1916

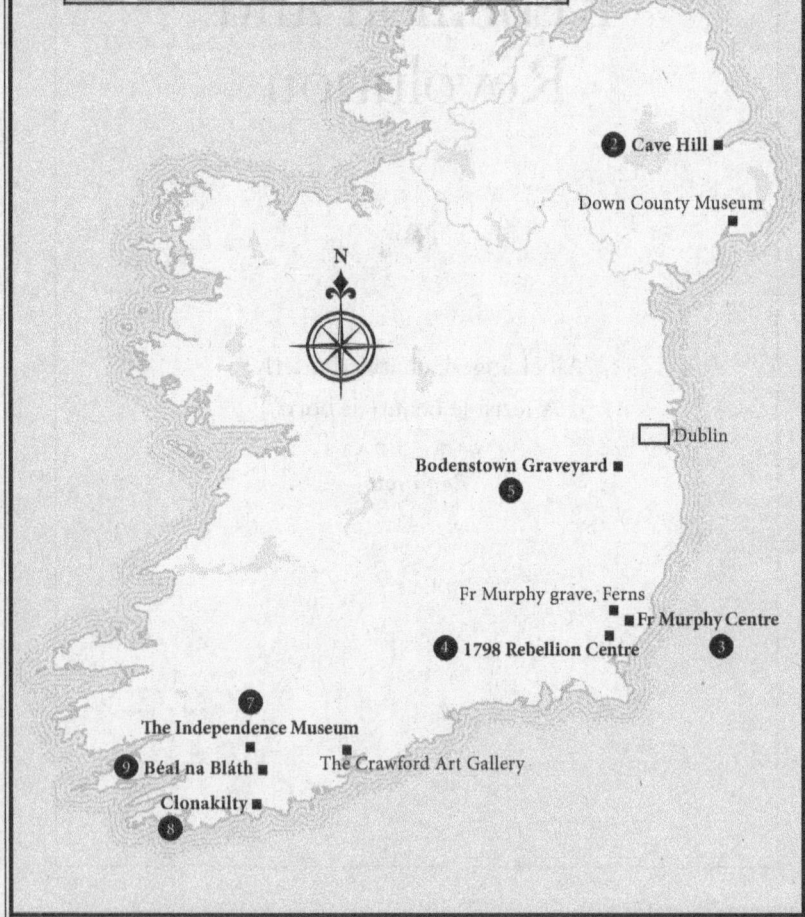

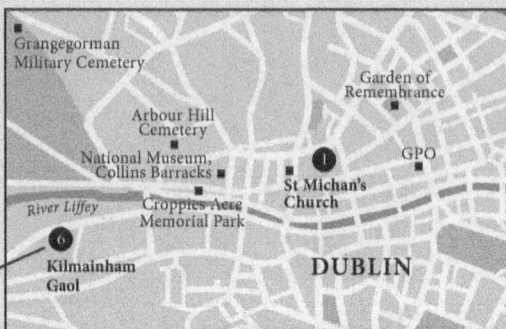

Grangegorman
Military Cemetery

Garden of
Remembrance

Arbour Hill
Cemetery

National Museum,
Collins Barracks

GPO

①

**St Michan's
Church**

River Liffey

Croppies Acre
Memorial Park

⑥

DUBLIN

**Kilmainham
Gaol**

② **Cave Hill** ■

Down County Museum ■

N

☐ Dublin

Bodenstown Graveyard ■

⑤

Fr Murphy grave, Ferns ■

■ **Fr Murphy Centre**

④ **1798 Rebellion Centre**

③

⑦

The Independence Museum ■

⑨ **Béal na Bláth** ■   The Crawford Art Gallery ■

Clonakilty ■

⑧

On a sunny May morning in Dublin I sat on a bench at Croppies Acre, a small park that sits between the National Museum and the River Liffey. Under my feet the bodies of hundreds of rebels, members of the Society of United Irishmen, are said to lie, buried there during the 1798 Rebellion, the largest rebellion ever to take place in Ireland. Over the course of four months, between the end of May and early September, members of the Society of United Irishmen and their supporters fought to break Ireland's connection with Britain. There are two memorials to the Rebellion in the park – one placed there to mark the centenary in 1898, and a more substantial one in 1998 (which currently operates as a windbreak for drug users) – though no evidence that any bodies are actually buried at Croppies Acre has ever been found.

I left the park and walked to nearby St Michan's Church. The church is best known for the mummified bodies in its crypt, though I was going to see a rather different object – the death mask of Theobald Wolfe Tone. I was embarking on a journey to sites associated with the 1798 Rebellion and I wanted to get as close as possible to the most fascinating and beguiling of all the United Irish leaders. For every rebellion, only a handful of individuals get national recognition, but at different sites a few local names are added to the role of honour – for the 1798 Rebellion the national hero Theobald Wolfe Tone is joined by Fr John Murphy in Boolavogue, Billy Byrne in Wicklow, Thomas Russell in Downpatrick and Tadhg an Astna O'Donovan in Clonakilty. But Tone is the constant. He almost single-handedly represents the 1798 Rebellion, which is quite an achievement for a man who took no active part in it.

In his writings – his diaries, his journalism, his correspondence – Tone was all things to all people, or at least to all republicans. He was both radical and revolutionary, but also accessible and fun. Alongside the politics, his diaries give a sense of a man who had 'a wild spirit of adventure'. In the diaries he is by turns reflective, bored, ebullient, drunk, amusing and polemical. Friends and enemies alike were often 'rascals' and 'scoundrels'. In March 1792, Tone was in Belfast, the place he felt most at home. A late-night diary entry reads: 'Dinner at the Donegall Arms. Everybody as happy as a king . . . Huzza! God bless everybody! . . . Huzza! Generally drunk. Broke my glass thumping the table. Home, God knows how or when. Huzza! . . . Sleep at last!' In March 1796, Tone was wandering Paris and missing his drinking companions: 'the French know how to be happy, or at least to be gay, better than all the world besides. The Irish come near them, but the Irish all drink . . . and the French are very sober. I live very soberly at present, having retrenched my quantity of wine one half.' The following day he noted: 'St Patrick's Day. Dined *alone* in the Champs Elysées [sic]. Sad! Sad!' As his biographer, Mari-anne Elliott, put it, 'Everything about his life – his intense loyalty to friends, his capacity for sacrifice, his inflated sense of honour, his rakish youth and dramatic death – is the very stuff of roman-tic legend.' When Thomas Davis and his fellow Young Irelanders came to construct their own version of Irish history in the mid-nineteenth century, Theobald Wolfe Tone was the embodiment of their heroic vision of Ireland's past.

Until the early twentieth century, death masks were often made of the famous, the notorious and the wealthy. Multiple copies were created and distributed among family and friends or put on show as a permanent three-dimensional reminder of the deceased – a dead hero or demon frozen in time. Very often, death masks were made by the doctor who first saw the body, for time is of the essence in such matters. The eyelids and lips were closed, the hair combed and oiled so that the plaster didn't stick to it. Then layers of plaster bandages were applied to the face and left to dry. When dried, the plaster mould was taken off and wax or metal poured into the mould to make the mask, a process that could be repeated multiple times. Tone's death mask shares a vault with the coffined bodies of Henry and John Sheares, United Irishmen and brothers who were executed in 1798. Peering past the coffins, I could see the outline of Tone's grey face propped up against the wall at the far end of the vault, but there was no sense of the vibrant, living, breathing Tone emanating from the spectral face that gazed blankly out at me. Perhaps beginning this trip underground, surrounded by the dead of the Rebellion, was a bad idea. I decided to go to where it all began.

Cave Hill is Belfast's iconic landmark, a great ridge of basalt that juts out over the city. From a distance its profile looks like the face of a sleeping giant, and it's believed to have been the inspiration for Jonathan Swift's *Gulliver's Travels*. One Friday morning I clambered up Cave Hill, passing some of the man-made caves which may have been rudimentary iron mines. At McArt's Fort – a ring fort perched at the highest point of the hill – I gazed down at the city spread out far below me. I had a rather crumpled copy of Andrew Nicholl's 1828 painting of the fort with me, and the differences in the landscape were striking. The hill remains as it was in Tone's day, but the vista below has been transformed. In the painting there's only one farmhouse amidst the patchwork of fields that flows all the way to the coast, while today urban sprawl has replaced every field with roads and housing estates. To my left, the coast meandered up towards Carrickfergus, where King William landed in 1690, and to my right, beyond the housing estates, Belfast city centre crouched low. There was a ferry beginning its journey to Cairnryan or Birkenhead; I could see the angular yellow limbs of Samson and Goliath, the towering Harland and Wolff gantry cranes, and in front of them, the sharp edges of Titanic Belfast glinting in the sun. Further right, I could see beyond Waterworks Park to Crumlin Road Gaol, and far beyond that the green oasis of the Botanic Gardens, home of the Ulster Museum. Many of the city's dark-tourism sights could be taken in at a glance, but very little of eighteenth-century Belfast, the United Irishmen's town, could be glimpsed from my vantage point. Much of it has been demolished and almost all of what remains is hidden by larger, newer buildings.

I had come to McArt's Fort because it was here that, in the summer of 1795, leaders of the United Irishmen, including Theobald Wolfe Tone, Samuel Neilson and Thomas Russell, met. The Society of United Irishmen was established in Belfast in 1791. Inspired by the ideals of the French Revolution, radical Presbyterians, Catholics and members of the Church of Ireland came together to plan for an independent Ireland where all men (women didn't really feature) would be equal, irrespective of religion. Such radical thought was not welcomed by the authorities and when Britain (and by default Ireland) went to war with revolutionary France in 1793 increasingly repressive measures were taken against the society: the editors of its newspaper, the *Northern Star*, were arrested on charges of seditious libel, soldiers attacked the homes and properties of Belfast radicals and, in May 1794, the society was officially banned. Forced underground, it was transformed into a secret, oath-bound organization committed to taking revolutionary action to secure Irish freedom. There had been enough pontificating in pubs and newspaper columns. Action was required.

Perhaps Tone and his comrades chose McArt's Fort for seclusion, for Tone needed to keep a low profile, as he had been linked to a plot to secure French aid for an Irish rebellion. In the spring of 1794 he had reached a deal with the Irish administration and agreed to leave the country in order to avoid prosecution. But he dallied and a year later he was still in Ireland, where the authorities were under increasing pressure to arrest him. A quiet meeting place seemed sensible, but Tone was a talented propagandist and a dramatic gesture needed a grand backdrop. Cave Hill is the closest thing Belfast has to a mountaintop, and it was there that Tone and the other United Irishmen swore an oath 'never to desist in our efforts until we had subverted the authority of England over our country and asserted her independence'. This was a declaration of war: the radical United Irishmen had become revolutionaries.

In the wake of their declaration, Tone finally did leave Ireland, and went in search of French aid for the planned revolution. He was, to some extent, pushing at an open door, for the French Directory (the revolutionary committee which governed France at the time) was already mulling over sending an expeditionary force to Ireland. By the autumn of 1796 a large fleet of sailors and soldiers was being readied for dispatch to bolster the forces of the United Irishmen. This was a serious invasion fleet, comprised of forty-five ships and fifteen thousand men, commanded by Lazare Hoche, one of the great military heroes of the French Revolution. In late December the fleet sailed from Brest in France to Bantry Bay. What could go wrong? The answer was: everything. Poor organization meant the fleet sailed weeks later than intended; poor communication ensured that the United Irishmen were not expecting it. For good measure, a storm blew most of the French ships dramatically off course. Tone, dressed in the uniform of a French adjutant-general (for he had joined the French Revolutionary Army), was on board the *Indomptable*, one of the ships that did manage to stagger into Bantry Bay. There he wrote in his diary that he was 'near enough to toss a biscuit ashore'. But thanks to a combination of indecision by the commanders who had reached Bantry Bay, bad weather and a scattered fleet, neither he nor any French soldier landed. 'England,' Tone noted, had 'not had such an escape since the Spanish Armada . . . Well, let me think no more about it; it is lost, and let it go!'

In response, the rattled Irish and British authorities implemented a series of brutal counter-insurgency measures throughout the country, but particularly in Ulster. For sixteen months, the armed forces wreaked havoc, burning houses, flogging, and 'pitchcapping', a practice in which tar was poured on the heads of suspected United Irishmen and then set alight. Arrests were frequent, charges less so, but as habeas corpus had been suspended, the arrested simply languished in prison without any prospect of release. Successful infiltration by spies also decimated the leadership and ranks of the United Irishmen, leaving their plans for rebellion hopelessly compromised. Bereft of commanders, lacking foreign aid and with sectarianism rising, the rebellion that took place in the summer of 1798 was haphazard and, from a United Irish perspective, disastrous.

The 1798 Rebellion has had at least three incarnations – the Rebellion itself, the centenary in 1898 and the bicentenary in 1998. In 1898, amid the surge in nationalist sentiment inspired by the Celtic Revival, it was celebrated as a 'faith and fatherland' rebellion, a Catholic fight for an Irish republic. That rather conveniently overlooked the fact that most of the United Irish leaders were Protestant (so much for the idea that to be properly Irish you had to be Catholic), that the Catholic Church had very limited involvement in the Rebellion (and indeed opposed it), and that the objective had never been to create a Catholic state. But by 1998 the Rebellion had been transformed from a nationalist and Catholic rebellion into one which highlighted the cooperation of Catholics and Protestants, united in a desire to create a more equitable Ireland. The bicentenary was neatly affixed to the Northern Ireland Peace Process, when it seemed timely to reflect upon the fact that two hundred years earlier the United Irishmen had called for 'a brotherhood of affection . . . a union of power among Irishmen of all religious persuasions'. The Taoiseach, Bertie Ahern, went so far as to claim that 'it is precisely because of its enduring relevance that 1798 has never truly passed out of politics and into history'. A safe and somewhat sanitized version of the 1798 Rebellion was offered for public consumption.

I was a student in 1998 and spent much of the year traipsing around the country visiting exhibitions, attending conferences and watching re-enactments – it was impossible to set foot in Wicklow or Wexford without running into pikemen about to re-enact some heroic last stand. For the bicentenary, the National Museum of Ireland and the Ulster Museum hosted substantial (and impressive) exhibitions: *Fellowship of Freedom* and *Up in Arms*. And so when I set out this time I expected that my pursuit of the 1798 Rebellion through museums and heritage sites would take some time, but I was wrong. Twenty years is a long time in the commemoration game and museums had moved on – the exhibition in the National Museum of Ireland had been replaced by an exhibition on the 1916 Rising and in the Ulster Museum by one on the Troubles. In many museums trace elements of the 1798 Rebellion can still be found, tucked away in glass cabinets or in a few sentences on panels, but for the most part the artefacts had been packed away and new rebellions and risings had filled the space. Today very little is said in museums about the complex identity politics associated with the United Irishmen, and in some communities the idea of northern Presbyterians as the first Irish republicans was (and perhaps still is) something to be ignored, not commemorated and certainly not celebrated. Indeed, the only museum which acknowledges this complexity is Down County Museum in Downpatrick, where a single panel notes that 'the story of '98 and the legacy of the United Irishmen has been debated and disputed for over two hundred years. It continues to generate discussion today.' It may 'generate discussion', but not in this or any other museum.

My grandfather, Denis, was a proud Wexford man, steeped in the Irish republican tradition – his father had fought in the Easter Rising in 1916, but like many others he learned his history through stories and songs. For Grandad, the ballad 'Boolavogue' was more than a ballad extolling the heroic actions of Fr Murphy, it was an accurate blow-by-blow account of Murphy's participation in the 1798 Rebellion. It made no difference to him that I'd spent years researching the history of Irish republicanism in the 1790s. Sure, what did I know? Tone was irrelevant. John Murphy was the only United Irishman that mattered; he was the great leader, everyone else a mere footnote in Fr Murphy's tragic, heroic tale. But Murphy's posthumous fame rests largely on a desire by Irish nationalists of the late nineteenth century for Catholic leaders of what they had recast as a Catholic nationalist rebellion. 'Boolavogue', written in 1898 by P. J. McCall, duly depicts Murphy as a courageous Catholic leader of a Catholic cause. It's a classic nationalist ballad, where initial victory is replaced by heroic failure followed by noble death. In the ballad Fr Murphy urges his men to 'Arm! Arm! . . . for Ireland's freedom we fight or die.' But their deaths were not in vain (they never are in these sort of ballads) and later generations would rise again for 'the causes that called you may call tomorrow/ In another fight for the green again'. 'Boolavogue' is a good example of the way in which ballads and poetry tend to be strong on emotion and weak on detail – songs often preserve the emotional impact of a rebellion at the expense of hard facts, or at least a nuanced understanding of the facts. I'm a fan of Boney M's 'Rasputin', but I wouldn't rely on it as a guide to the Russian Revolution.

There is a visitor centre devoted to Fr Murphy at Boolavogue and (despite groans of anguish from Al, who for some reason appeared not to relish hearing rebel songs, even though I explained to him that this was an important element in his atonement for the crimes of his ancestors) it seemed fitting to play 'Boolavogue' as we headed off to visit it. The Fr Murphy Centre got a big musical build-up, but when we arrived one Sunday afternoon the empty car park did not bode well. We bought our tickets and were directed through the coffee shop to the farm buildings beyond. The buildings, arranged in a square, have been lovingly restored. The problem was what was – or wasn't – inside. The farmhouse at the core of the site has a very tangential link to Fr Murphy – he stayed there, but it wasn't his home. There's not very much about the Rebellion at the centre, though we did learn a lot about local folklore and country life – I was particularly taken by the plough Ned Jones used to win the inaugural ploughing championship in 1931.

Perhaps on busier days there are guides who explain Fr Murphy's role. I hope so, for he did play an interesting (if far from pivotal) part in the Rebellion. At the outbreak, not only was he not actively involved, but he cautioned his parishioners against taking part. However, after witnessing a brutal assault by troops on some locals he had a change of heart and threw himself into the fight, taking part in the Battle of Oulart Hill then moving on to action in the nearby town of Enniscorthy. He escaped un-scathed from the Battle of Vinegar Hill, which signalled the defeat of the rebels in Enniscorthy, but was later captured near Carlow and charged with treason. He was found guilty and hanged, and after his death decapitated, his head put on a spike and his body burned in a tar barrel. We left Boolavogue behind and headed further south, pausing in Ferns at his grave, where a small and somewhat macabre plaque notes that 'his part re-mains are buried here'.

From Ferns we travelled to Enniscorthy, where I left Al to ramble around the town while I visited the 1798 Rebellion Centre. The centre is a little bit like a 3D academic essay where a very earnest student outlines the causes of the Rebellion, describes the Rebellion itself and analyses the consequences. Visitors can trace a political genealogy from the 1798 Rebellion to everything from the repeal movement of the 1830s to the Ulster Volunteer Force (UVF) decommissioning arms in 2009. There's a barrage of information about the Age of Enlightenment, and much about the political philosophers Edmund Burke and Thomas Paine going toe to toe with their treatises. Burke's *Reflections on the Revolution in France* was an influential critique of the French Revolution, while Paine's *The Rights of Man* defended its values – liberty, fraternity, equality – and was one of the inspirations for the United Irishmen. Pages from *The Rights of Man* flutter down one wall, while in another room Tone, Prime Minister William Pitt and others are represented as giant plastic chess pieces arranged on a large chessboard. Alongside the chess figures, explanatory text (I use the term loosely) reads:

As Revolutionaries and Counter-Revolutionaries manoeuvred for advantage, the contest between them became a deadly game of strategy . . . The Counter-Revolution King (the British Government) is under no direct threat. In contrast, the Revolution King (the United Irishmen) is in check from a castle (the Anglo-Irish Ascendancy) and must move away from this attack. In the middle of the board two pawns (the Orange Order and the Defenders) threaten each other. If the Counter-Revolution pawn captures the Revolution pawn, the Knight (the French) can recapture and will then be attacking the Bishop (the Irish Catholic Church).

As attempts to demystify the conflict go, it's quite astonishingly inept. They might as well have used the musical *Chess*, with Barbara Dickson as the Anglo-Irish Ascendancy and Elaine Paige as Tone.

Things improved for me, if not for the rebels, when I reached the section on the Battle of Vinegar Hill, which took place on 21 June. I walked into a large, dark, cavernous space and up a slight incline to what seemed, in the gloom, like a viewing platform. It was somewhat unnerving, as I was the only live person there, though plenty of mannequins kept me company. A pikeman (or 'croppy', as the rebels were often called, as they kept their hair short) stood beside me, jabbing his pike into the chest of one of the redcoats while a wounded rebel slumped over the rail in front of me. Suddenly a cacophony exploded all around. Military orders were barked amid the blasts of cannon and musket fire. Against the back wall a film of the battle played and a narrator told the story of Vinegar Hill. Shouts and screams echoed through the space while smoke curled around my legs and bright flashes briefly illuminated parts of the 'hillside' strewn with the broken bodies of rebels and soldiers. A young boy ranged across the hill in search of rebel family members until, in a haze of dry ice and strobe lighting, he was killed. The battle ended and I emerged, rather shaken, beside a plastic gravestone representing a dead rebel, one of the hundreds killed that day. While it's impossible to convey the reality of battle in a museum, the centre gives it a commendable stab (no pun intended).

Leaving the 1798 Centre, I walked through Enniscorthy, and along the River Slaney en route to the real Vinegar Hill. I thought

of Seamus Heaney's 'Requiem for the Croppies', which seemed
particularly apt that May morning:

> Until, on Vinegar Hill, the fatal conclave.
> Terraced thousands died, shaking scythes at cannon.
> The hillside blushed, soaked in our broken wave.
> They buried us without shroud or coffin
> And in August the barley grew up out of the grave.

The death toll that day was in the hundreds, not the thousands (again, poetry – weak on precision), but over the course of
the Rebellion tens of thousands did die. The seeds had been
sown on Cave Hill, but the harvest that was reaped on Vinegar
Hill was a very bloody one.

The Battle of Vinegar Hill sounded the death knell for the
Rebellion. In August, three French ships and 1,100 soldiers
landed near Killala in County Mayo, but they were swiftly
defeated. Tone, who was back in France trying to secure another
invasion force, was unaware of the outbreak of rebellion, but as
soon as he learned of it he was determined to return. By October
1798 he was on board the *Hoche*, one of nine French ships bound
for Ireland. But British spies tracked the movement of the little
fleet from the moment it set sail, and as it neared the Donegal
coast the Royal Navy attacked. The *Hoche* was captured and
Tone was recognized, arrested and taken to Dublin, where he
was court-martialled and imprisoned in the Royal Barracks
(now Collins Barracks and home to the National Museum of
Ireland). Tone was found guilty of treason and sentenced to be

hanged, drawn and quartered, but the death sentence was never carried out because Tone cut his own throat, and died of the wound on 19 November 1798. This was not the story that circulated until well into the twentieth century, which was that Tone had been murdered in jail by an English soldier. If Tone's Protestantism was an inconvenient fact for Catholic Ireland, his suicide was unmentionable.

Tone was buried in Bodenstown graveyard, near Sallins, County Kildare, close to the Tone family home. For many years after his death the only visitors were locals, but that all changed when Thomas Davis, a Young Irelander, decided that Tone was the embodiment of Irish republicanism. Davis's poem 'Tone's Grave', published in *The Nation* newspaper in the 1840s, bemoaned the fact that Tone had been almost forgotten: 'A martyr for Ireland – his grave has no stone,/ His name seldom named, and his virtues unknown.' The Young Irelanders set about correcting that, moulding Tone into the father of Irish republicanism and casting themselves as his heirs. From the 1870s, Tone's grave became a site of annual pilgrimage for every variety of nationalist and republican, all of whom wanted to call themselves his heir and have him legitimize their cause. This was not hard. Tone was romantic, swashbuckling and charismatic, and he wrote so much that there's something for everyone to latch on to, whether it is endorsing the use of violence or championing parliamentary reform. As the writer Séan Ó Faoláin observed, Tone 'started much. Without him Republicanism in Ireland would virtually have no tradition.' At a speech at Tone's grave in 1913, Patrick Pearse proclaimed him 'the greatest of all Irish nationalists' and his grave 'the holiest ground in Ireland'. A small rostrum has been constructed at the graveyard and there's a bronze plaque with Tone's image engraved on it. On the day I visited there were no large crowds, just me and my friend Peter, but obviously that didn't stop me from climbing the steps to the platform and declaiming my own words of wisdom to my audience of one.

The Young Irelanders themselves have left little trace – just a sword, and the farmhouse near Ballingarry, County Tipperary, where their brief rebellion took place. Words, not deeds, were their forte – not sufficient fodder for a successful revolution. The Young Irelanders like Davis, John Mitchel, John O'Mahony, William Smith O'Brien and Thomas Francis Meagher were cultural nationalists heavily influenced by European romantic nationalism. They were passionate about promoting Irish literature, history and language. They established *The Nation* in 1842 and through its pages created a new, nationalist version of Irish history. Politically, for a time, they allied themselves to Daniel O'Connell's repeal movement, which was determined to reverse the Act of Union, which had been passed in 1800, largely as a reaction to the 1798 Rebellion, and restore the Irish Parliament. However, they split with O'Connell in July 1846, largely because he refused to countenance the use of force. Meagher, known by some as 'Meagher of the Sword', broke the Young Ireland link with O'Connell when he made a speech announcing: 'I do not abhor the use of arms in the vindication of national rights. There are times when arms will alone suffice . . . force must be used against force . . . The man that will listen to reason, let him be reasoned with; but it is the weaponed arm of the patriot that can alone avail . . . against battalioned despots.' Meagher's sword was not simply metaphorical – it can be seen in the Bishop's Palace in Waterford (though the sword on display is connected to his later military career in the United States, rather than his rebel past in Ireland. Indeed, it is a curious fact that the Bishop's Palace is one of the few places in the country where there is any mention of the American Civil War – a war in which over two hundred thousand Irishmen fought).

Not content with rewriting the history of Ireland, some Young Irelanders, like Meagher and Smith O'Brien, wanted to take action, inspired by the wave of revolutions that engulfed much of Europe in 1848. But trying to start a revolution in a country in the midst of a devastating famine proved impossible. While the 1848 rebellions toppled governments from Paris to Palermo, the Irish attempt to overthrow the status quo in July of that year was rightly given the Swiftian title the 'Battle of Widow McCormack's Cabbage Patch'. In both the months preceding and following the botched rebellion, most of the senior Young Irelanders were arrested, found guilty of treason felony and sentenced to transportation for life (the sentence for treason was death; for treason felony, the prisoner could be transported). Many of them, including Mitchel, O'Mahony and Meagher, forged successful careers in America and, most importantly for their legacy, created a new history of Ireland which enabled them to neatly link past rebellions to their own actions in 1848 and to the new Fenian movement of the 1860s. The Young Irelanders thus helped to forge an Irish identity based around a

romanticized version of the past, of myth and legend, of literature and poetry, the narrative of heroic failure which has
remained the standard narrative for over 150 years.

Two veterans of the Young Ireland movement, James Stephens
and John O'Mahony, were behind the establishment of the
Irish Republican Brotherhood (IRB) and the Fenians. Stephens
founded the IRB in Dublin in 1858, and the following year
the Fenians were established in New York by O'Mahony.
These sister organizations, often collectively referred to as 'the
Fenians', were the first Irish republican groups established
with the primary intention of using force to achieve Irish
freedom. Much of their activity was focussed on the US, Canada and Britain, and so, like the Young Irelanders, they make
only fleeting appearances in Irish museums, but they do have
a presence in prison sites and cemeteries. The most famous
site associated with the Fenians is not a battlefield, but a burial
plot, and it's renowned not so much for the man buried beneath the soil but for the speech that was made at his
funeral.

Jeremiah O'Donovan Rossa was a member of the IRB. When
first established, the intention was that the IRB would devise
and carry out attacks on British control of Ireland, while the
Fenians based in the United States would fund such efforts. The
reality wasn't quite so clear cut. In 1865 O'Donovan Rossa was
arrested because of his role as journalist and business manager
for the IRB newspaper, *The Irish People*, and was put on trial for
high treason. He was found guilty and sentenced to penal

servitude for life. During his six years in prison he was subjected to torture, including one period when his hands were kept cuffed behind his back for thirty-five days. In 1871 he was released on condition that he leave Ireland. Rossa sailed for America, where he became a senior figure in the Fenians and organized the 'skirmishing campaign', which attempted to spread terror in Britain though a series of bomb blasts in public places. However, by the time he died in New York in January 1915, he was largely forgotten. His funeral would change all that. In Irish republican circles, death is good for your reputation. Rossa's body was returned to Ireland and he was buried with great pageantry in Dublin's Glasnevin Cemetery, an act which ensured his place in the pantheon of Irish republican heroes. His was far from the first funeral to be used for political purposes, but it was impressively stage-managed by the veteran Fenian Tom Clarke, who invited the firebrand Patrick Pearse to deliver the graveside oration. Revolution and religion are frequently intertwined, and nowhere is this more apparent than at graveside orations where the revolutionary orator usurps the place of the priest.

Visitors to Glasnevin Cemetery can relive the O'Donovan Rossa burial every day at 2.30 p.m., when an actor dressed in the green livery of an Irish Volunteer delivers Pearse's deliberately provocative and rousing speech. I've watched this performance several times and I find it really uncomfortable. It's not part of a tour. It takes place very close to the crematorium, the shop and the café, spaces frequented by many people who have no intention of taking a tour or visiting the museum. Out of context, the event looks and sounds bizarre – a man in military uniform essentially declaring war on Britain. For those who know about Patrick Pearse and the 1916 Rising, it all makes sense, but I worry about what passers-by might make of the speech as it reaches its rousing crescendo: 'The Fools, the Fools, the Fools! – they have left us our Fenian dead – And while Ireland holds these graves, Ireland unfree shall never be at peace.' I hope the re-enactment doesn't have quite the impact the original did, for in twenty-six words Pearse not only redeemed Rossa's reputation but inspired a revolution.

The envelopes were pale cream and satisfyingly weighty, the sort that might contain a wedding invitation. Inside one was a gilt-edged card complete with an embossed harp. It was from the Taoiseach, Enda Kenny, inviting me to attend a ceremony to mark the centenary of the Easter Rising – an insurrection which, although swiftly crushed, marked the beginning of a journey that culminated in the establishment of a twenty-six-county Irish Free State in 1922. The second envelope contained another invitation, this one to the 'Relatives' Event' at the RDS. My great-grandfather, Jack O'Brien, had joined the Irish Volunteers (a paramilitary force founded in November 1913) in February 1914 and fought in the Easter Rising in County Wexford. During Easter week he attacked the Royal Irish Constabulary (RIC) barracks, destroyed telegraph lines in Enniscorthy and held up trains travelling along the east coast, before moving to Ferns, which he and others held until the surrender. He was arrested and taken to Lewes Jail, a few miles from Brighton in East Sussex, and from there he, along with 1,800 other prisoners, was sent to Frongoch Camp in Wales, where he remained until the camp was closed and all prisoners were released, just before Christmas 1916.

As a child, I heard that Grandfather O'Brien had fought in the Rising, but then so had everyone, it seemed. He died before I was born and, apparently, he never spoke of his involvement, yet every Easter he flew the tricolour in his garden in Athlone. As I was growing up, the print he owned of Michael Collins hung in our house – reflecting family ties rather than an affiliation to any political party. My grandfather was not alone in his reluctance to discuss the past. For decades afterwards, the 1916 Rising was highly contentious. How should it be remembered? It seemed as if successive Irish governments were fearful of making too much of the Easter Rising in case it provoked trouble in the North. It was fifty years before the Garden of Remembrance, on Parnell Square in Dublin, was opened – a project that had begun back in the thirties – and even then it was dedicated to all those who gave their lives in the cause of Irish freedom, not just in the Easter Rising. Between 1972 and 2006 the Irish government suspended its annual military parade past the GPO, fearful that it sent the wrong signal during the Troubles, but that only enabled Sinn Féin and the IRA in its various guises to take ownership of the Rising and claim to be its heirs. There's nothing like being afraid of your own history. But by 2016, nearly twenty years after the Good Friday Agreement was signed, it seemed that the country was finally ready to embrace the 1916 Rising.

Several years ago, my father gave me my great-grandfather's medals. These and a photo of him wearing them in 1966 hang above my desk, reminding me daily of his story. When I was a child, his exploits had never captured my imagination – I think this was partly because he had fought in County Wexford and the focus of books, commemorations and history lessons was always Dublin: everything that happened outside the city was a footnote, if mentioned at all. I'm very proud that my great-grandfather took part in the Rising, but I can claim neither credit nor blame for any of his actions. The men and women who fought in 1916 did so in the hope of creating an Irish republic, a place where, at least on paper, everyone would be equal. A place where there would be no aristocracy, where a blood link conferred no special status, where positions weren't inherited by accident of birth. The outcome was, of course, quite different. Eldest sons inherited lands, daughters inherited little. Afterwards, a small number of families (many connected with the 1916 Rising) came to dominate Irish political, economic and cultural life. Holding an event where simply being related to someone guaranteed an invitation seemed anathema to the memory of the revolution.

And yet, despite my reservations, I went along to the Relatives' Event. As requested by the organizers, my father and I had sent in photos of my great-grandfather and dutifully watched the giant screens as photos of those who'd fought scrolled across them, then scrolled across again, and again. Jack O'Brien never appeared. I was a little put out. The slideshow was followed by an impressive and moving speech by President Michael D. Higgins in which he remembered the 'sung and unsung heroes of 1916' and the 'sacrifices made by so many of those who helped to build our nation'. But in a speech that lasted over twenty minutes and mentioned Dublin and its suburbs numerous times (seventeen, in fact – I was counting), just one line acknowledged that 'our road to independence stretched too, of course, beyond the capital city . . . '. Of course, but let's not talk about that. And so all those who fought outside the capital remain largely forgotten. The government may tell a more inclusive and less partial story these days, but it still shapes the official narrative into neat little (sound) bite-sized morsels that leave little room for subtlety, complexity, or events 'beyond the capital city'. It's the job of museums to offer more nuanced and sophisticated narratives and of historians to counter the politicians' impulse to oversimplify.

On Easter Monday 2016, over 750,000 people crowded the streets of Dublin during an event named 'Reflecting the Rising', guided to a plethora of public talks, exhibitions, concerts and performances by volunteers dressed in period costume. These events were commemorating the fact that one hundred years earlier, on 24 April 1916, just over 150 Irish republicans took over the General Post Office on Dublin's O'Connell Street and proclaimed a republic. The Proclamation was signed by the seven men of the Military Council of the IRB who had planned the Rising, though it's thought to have been largely written by Patrick Pearse and James Connolly. The Proclamation declared the 'right of the people of Ireland to the ownership of Ireland', promised 'religious and civil liberty, equal rights and equal opportunities to all its citizens', and resolved 'to pursue the happiness and prosperity of the whole nation and of all its parts, cherishing all the children of the nation equally'.

Across the city centre, between the two canals, the rebels

briefly secured a number of outposts, from Boland's Mills in the east, commanded by Éamon de Valera, to St Stephen's Green and the Jameson's Distillery on Marrowbone Lane in the south and the GPO on the north side of the Liffey. Outside Dublin, the main rebel activity took place in Galway, Wexford and Meath. One bloody week later it was all over. Nearly 500 people had been killed and over 2,500 wounded (most of them civilians) and much of central Dublin had been destroyed by gunfire, shelling, fire and arson. The Rising had not received popular backing – many people had family members fighting in the British Army at the time, huge swathes of the city had been destroyed by bombardment and fires, shops had been looted and jobs lost. As the defeated rebels were marched into custody, crowds shouted insults and threw rotten fruit and vegetables.

There are two iconic sites associated with the Easter Rising: the General Post Office, which was the rebel headquarters for most of the Rising, and Kilmainham Gaol, where many of those who fought were imprisoned and where fourteen of the sixteen executed rebels were shot by firing squad. Designed by Francis Johnson, the GPO opened its doors in 1818, a three-storey neo-classical building that dominates O'Connell Street (then Sackville Street). It's undoubtedly the most impressive place to purchase a stamp in Ireland. Colm Tóibín has written that on a busy day 'it is possible to go into the General Post Office . . . to post a letter or buy a TV licence and not think at all about the 1916 Rebellion'. Not for me. Whenever I approach the building I start to look for the bullet holes in the columns. All the columns are pockmarked and scarred, no doubt in part because they've stood there for two hundred years, but as a child I was happy to believe that every indentation came from a bullet fired in April 1916. Inside, I imagine the smell of cordite and the sound of falling plaster and I think of Walter Paget's painting *Birth of an Irish Republic*, a copy of which hung on the wall of my great-grandmother's living room in Athlone. Paget wasn't in the GPO during the Rising – indeed, he may never have been in Ireland – but his evocative painting quickly became iconic. In it, a wounded James Connolly lies on a stretcher. Patrick Pearse stands beside him with pistol drawn. All around them there is mayhem as Volunteers tend to the wounded, try to put out fires and crowd around the windows, firing into the street.

The round-headed window from which Paget's Volunteers shot is now occupied by Oliver Sheppard's bronze statue of the death of the mythic warrior Cúchulainn, which was unveiled as a memorial to the Rising in 1935. Today, the interior of the post office is all brass and highly polished wood; during the Rising, fire wreaked such havoc that only the shell of the building survived. And yet every time I queue to buy some stamps I wonder how customers like me felt on Easter Monday morning 1916, when members of the Irish Volunteers and Irish Citizen Army led by Pearse and Connolly marched in and began their fight for an Irish republic.

Now, as well as operating as the post office, the building is home to GPO Witness History (a name that must be meaningless to any visitors who don't already know what the GPO signifies in Irish history). In the museum, I hoped to get a better sense of what it was like to be in the building as Easter week unfolded. But, rather disappointingly for a site-specific museum, almost all traces of the site have been eradicated. Housed in the basement, the exhibition makes little of its location and, while it's full of informative text panels, artefacts and all sorts of technological wonders, I was hardly aware that I was standing in the eye of the 1916 storm. I could have been anywhere. Of the twenty-one zones in the museum, only two – 'In the GPO' and 'From Ruin to Radio' – describe what happened in the building and, although there is an engaging film which shows the Rising from the perspective of the GPO, the story of the building and the Rising was never brought to life. I wanted to know what could be seen from the windows, where the leaders met and strategized, what happened when the building was aflame.

I found one exhibit particularly moving. In the corner of one display case lies an open autograph book with a black-and-white ink drawing of a dog with a mournful expression. Underneath, the artist has written 'Fed up with the War'. The war in question was the First World War. It was drawn on 19 April 1916, just five days before the Easter Rising began and seven days before the owner of this little autograph book was killed. Thirteen-year-old Madge Veale lived at 103 Haddington Road and had been given the notebook as a Christmas present. She'd spent the first four months of 1916 getting friends and family to write in it. On 26 April, Madge was shot and killed at her window, peering out at a gun battle between the Irish Volunteers and the Sherwood Foresters, a British Army infantry regiment, on the street below. She was one of forty children killed during the Easter Rising.

The National Museum has more than fifteen thousand objects

in its Easter Week collection and three hundred of their finest and most interesting pieces were displayed in the *Proclaiming a Republic* exhibition which ran from Easter 2016 to Easter 2020. Many of these everyday objects were rendered significant by their association with individuals – Tom Clarke's razor, Patrick Pearse's glasses. Two in particular caught my eye – a stack of sugar bowls fused together by the heat of the fire that tore through the Metropole Hotel on O'Connell Street during the Rising, and 'Dr McKenzie's Catarrh-Cure Smelling Bottle' – a beautiful emerald-green glass bottle that still contained some smelling-salts which were, allegedly, the 'best remedy for faintness and dizziness'. The bottle was used by Eilís O'Connell, a member of Cumman na mBan (the women's wing of the Irish Volunteers), when she was treating wounded rebels at the makeshift hospital at Father Mathew Hall on Church Street. It's a touching reminder of the brave medical workers who are inevitably caught up in conflict and of the thousands of individual stories of injuries and death that are all too often overlooked in tales of battles and wars. The day after I first saw this bottle I stumbled across another one for sale at a flea market and bought it. I keep it in my 'cabinet of curiosities', a small, black lacquered cabinet decorated with mother-of-pearl birds and flowers that used to live in a now-closed convent. On my travels, I picked up many such mementoes and over time it grew into a substantial collection.

Kilmainham Gaol opened in 1796 and was operational almost continuously until 1924. Over the course of almost 130 years, thousands of prisoners passed under the five writhing serpents carved into the stone above the front door. The snakes are entwined in chains and have been said to represent five serious crimes – murder, rape, theft, treason and piracy – crimes for which some prisoners were hanged from the gallows that once protruded above the entrance. Though most of the prisoners held in Kilmainham were not political prisoners, a Who's Who of Irish political prisoners did occupy the cells – from the United Irishmen of the 1790s to the leaders of the Easter Rising (though many of them were not famous at the time) – and every tour visits the narrow, cold, damp '1916 Corridor' in the jail's west wing. Although 1916 prisoners were scattered throughout the jail, at least six of the fourteen men executed in the prison were held on this corridor – Tom Clarke, Patrick Pearse, Joseph Plunkett, Willie Pearse, Michael Malin and Thomas MacDonagh. Visitors shuffle along a narrow wooden walkway, some fanning out to stand along the short bridges that link it to the cells (there are iron railings on the side, and wire mesh, protecting visitors from the sheer drop to the flagstones twenty foot below). Small name tags above the doors indicate who was held in the cells. The guide tells the tour group about the wedding of Joseph Plunkett and Grace Gifford, which took place in the prison chapel just hours before Plunkett was executed, and it reminds me of Margaret Pearse's story of being taken to Kilmainham just before dawn on 3 May to see her brother Patrick before his execution. She was escorted to the prison by two army officers. 'Just as we came to the middle of the road,' she recalled, 'we heard a volley. And they looked at each other and one said, "We're late."'

From the '1916 Corridor', tour groups are taken outside through the exercise yards into what was once the stone-breakers' yard, where men spent laborious hours every day. The huts that lined the space are gone, though you can still see their outline against the perimeter wall. The yard is empty except for a flagpole, two small black wooden crosses and a brass plaque affixed to a wall. Every tour group crunches its way across the gravel and assembles in a semicircle close to where, between 3 and 12 May 1916, fourteen of the men who took part in the Rising, including all seven signatories of the Proclamation, were executed. Thirteen were shot where one cross now stands; the fourteenth man, James Connolly, was executed at the second cross, close to the large wooden gates in the left-hand corner of the yard. Connolly had been badly wounded during the Rising and had been taken to the temporary hospital at Dublin Castle, where he was treated for his wound, court-martialled and sentenced to death. Early on the morning of 12 May he was taken by ambulance to Kilmainham Gaol. The ambulance drove into the stone-breakers' yard, and Connolly was taken out and strapped to a chair before the firing squad took aim and killed him. The executions, together with the arrest of some 3,500 people (far more than were involved in the Rising), shocked the nation, and public opinion about the Rising, to quote W. B. Yeats, 'changed utterly'. From a British perspective, court-martialling and executing the leaders of an armed insurrection who had sought and received support from Germany in the middle of the First World War may have been an obvious thing to do, but it was a disastrous decision which began a chain of events that spelled the beginning of the end for the British occupation of most of Ireland, and accelerated the collapse of the British Empire.

James Dillon, the deputy leader of the moderate Irish Parliamentary Party, was correct when, in a speech in the House of Commons the day before Connolly's execution, he railed against the government, who, he argued, were 'letting loose a river of blood . . . It is the first rebellion that ever took place in Ireland where you had a majority on your side . . . and now you are washing out our whole life work in a sea of blood.'

I followed the final journey of the executed men up to Arbour Hill Cemetery on the north side of the city, where they were buried in a single quicklime-lined grave in a corner of the parade ground of Arbour Hill Prison. There's something incongruous about the gravesite of the fourteen dead patriots, flanked as they are on one side by the bodies of British military men and their families and on the other by the sex offenders imprisoned in the jail. The executed men remained largely unvisited until, in 1955, that part of the prison site was opened to the public and a memorial area was developed. Today, a high limestone wall stands behind the burial plot. Engraved on it is a gilded cross and the words of the Proclamation in both Irish and English. It's a solemn, respectful memorial, but it does feel tucked away, almost a footnote to the main text that is Glasnevin Cemetery, where a broad sweep of dead Irish heroes are buried.

Just two kilometres away there's another cemetery which holds the bodies of men killed during the Easter Rising. On Blackhorse Avenue, very close to Dublin Zoo, is Grangegorman Military Cemetery, which opened in 1876. Alongside the graves of over six hundred men who died during the First World War are the bodies of soldiers serving in the British Army – many of them Irish – who were killed in the 1916 Rising. Their graves are just as well tended as those of the leaders of the 1916 Rising, if not as often visited.

The War of Independence, fought between Irish Republicans and forces of the British Crown in Ireland, began in January 1919 and ended in 1921. Since castles were obsolete by the early twentieth century, there was not a siege to be had; instead, the War of Independence was a guerrilla war which spread fear, brutality and destruction throughout the island. Despite the fact that Sinn Féin had not been actively involved in the 1916 Rising, in the wake of the rebellion many of those who had participated joined the party. The party fought the 1918 election on the basis that newly elected MPs would refuse to take their seats in the parliament at Westminster, that a republic would be declared and that Ireland would secede from the British Empire. It was also heavily involved in a popular campaign against the imposition of conscription in Ireland. The retrospective links to the Easter Rising and the anti-conscription campaign were a success, and in December 1918 Sinn Féin won 73 of the 105 Irish seats. An Irish Republic was duly declared, and on 9 January 1919 the first Dáil met in Dublin (despite the fact that many of the new TDs were still in prison). Coincidentally, on the same day, two RIC men were shot dead at Soloheadbeg in County Tipperary – deaths that are often regarded as the first casualties in the War of Independence.

In 1920 the number of attacks by the Irish Republican Army (as the Irish Volunteers were now called) intensified, and many rural RIC barracks were abandoned as the police force sought refuge in towns. In the local elections which took place in the summer of that year, Sinn Féin triumphed across most of the country and took over many of the functions of government, including tax collection, policing and running their own courts (the British system ran alongside the new Irish one, but was widely ignored). Paramilitary police were dispatched from Britain – Auxiliaries (members of the Auxiliary Division of the RIC who were generally former military officers who had served in the First World War) and the Black and Tans (so called because they wore a combination of a dark police uniform and military khaki). They behaved with great brutality, carrying out reprisals on civilians for IRA attacks and setting fire to farms, houses and towns.

For the most part, the War of Independence was, as Joost Augusteijn noted in the *Atlas of the Irish Revolution*, a conflict that was 'not a consequence of careful planning or design, but a result of a mixture of coincidence, unintended outcomes and local initiatives'. At the Independence Museum in Kilmurry, a small village ten kilometres from Macroom in County Cork, many local tales associated with the war are told. The museum had a wealth of stories to choose from, for, as historian John Borgonovo has observed, 'from 1919 to 1921, County Cork was the most violent county in Ireland'. I would have bypassed Kilmurry entirely but for one thing: Deirdre Bourke had opened the museum that afternoon just for me, and it was certainly worth the trip. Although the museum takes the long view of the fight for Irish independence, the spine of the collection relates to the War of Independence. There's an abundance of artefacts, including my old favourites – a Cromwellian cannonball and a famine soup pot, two items that appear in Irish museums so often I wonder if they get handed out as part of a starter pack whenever anyone decides they're going to open a new museum. But there are other, rarer, objects on display too – some branches roughly fashioned into replica handguns and used to train local Volunteers, and a beautifully embroidered Celtic Revival dress worn by Kate O'Callaghan, who was elected as a Sinn Féin TD for Limerick in 1921. Indeed, the great strength of the museum is that it tells the national story of the fight for Irish independence but roots it firmly in the local, with almost every artefact on display closely connected with County Cork. There's prison art created from animal bones, slate tiles and spent shells made by West Cork men held in prisons and internment camps during 1916 and the War of Independence. A number of local ambushes get a lot of attention, particularly that of Lissarda, just down the road, in August 1920, and the more famous one at Kilmichael (about twelve kilometres from the museum), which took place that November when a group of IRA men commanded by Tom Barry attacked and killed seventeen Auxiliaries. The rusting remains of a wheel salvaged from the wreckage of a Crossley Tender which had been transporting the Auxiliaries is on display. It's a peculiar trophy, reminding me that so often in war a seemingly innocuous object becomes infused with meaning because of its association with bloodshed.

Much of the collection relates to the MacSwiney family, and indeed the museum when it first opened in 1965 was called the

'Terence MacSwiney Memorial Museum' and was housed in the former home of MacSwiney's grandparents. MacSwiney had been Commandant of Cork No. 1 Brigade of the IRA and, when he was arrested in August 1920, he was both a TD and Lord Mayor of Cork. He was charged with sedition, found guilty and sentenced to two years' imprisonment in Brixton Prison in London. In protest, he went on hunger strike, and died on 25 October, the seventy-fourth day of his strike. There's a range of artefacts relating to MacSwiney, his wife Muriel, and his sister Mary (who was an active member of Cumman na mBan and a Sinn Féin TD in the twenties), including photographs, books, a lock of his hair and some wedding gifts. MacSwiney's funeral saw up to a hundred thousand people line the streets of Cork as his body was taken from the Catholic cathedral, across the city to be buried in the republican plot at St Finbarr's Cemetery; the front carriage wheel from the hearse that carried his coffin is on display. Six weeks after MacSwiney's burial, much of Cork city centre was set alight by Crown forces and destroyed. Over the following six months, around a thousand people were killed in the conflict (half of these in County Cork). By the summer of 1921, the IRA were short of men, arms and ammunition, but the British forces saw no way to comprehensively defeat them, and on 11 July 1921 a truce was announced.

Just how remarkable it was that a patchwork army had fought the British Empire to a standstill was brought home to me as I stood in the Crawford Gallery in Cork looking at Seán Keating's iconic *Men of the South*, which shows six men of the North Cork Brigade of the IRA with their rifles and bandoliers and mismatched assortment of suits and hats. Looking at this painting, it seemed almost impossible that such a rag-tag band of rebels had forced the enemy to the negotiating table. Many members of the IRA saw the truce as an opportunity to restock their arsenal and recruit more members. But the truce was no temporary end. In October, an Irish delegation led by Arthur Griffith (the Minister for Foreign Affairs) and Michael Collins (the Minister for Finance) travelled to London, and on 6 December 1921 the Anglo-Irish Treaty was signed. The treaty established a twenty-six-county Free State; the RIC was disbanded and the military garrison withdrawn, but the British Navy retained control of three naval bases. The treaty also confirmed the partition of Ireland, which had been initiated in 1920 under the Government of Ireland Act, but promised that an independent Boundary Commission would reassess the border. Collins argued that the treaty gave Ireland 'freedom – not the ultimate freedom which all nations hope for and struggle for, but freedom to achieve that end'. Not everyone agreed – indeed, the majority of the IRA was unhappy with the deal – and despite the fact that the Dáil narrowly voted in favour of the treaty (64 to 57) in January 1922, a slow march towards civil war had begun.

If Wolfe Tone personifies the 1798 Rebellion, then Michael Collins dominates the story of the Irish Revolution. In the space of just over six years he went from being a largely unknown Volunteer during the Rising to Minister for Finance in the First Dáil and, more memorably, Director of Organization and Intelligence. Following the ratification of the Anglo-Irish Treaty, he became Chairman of the Provisional Government of the Irish Free State and Commander-in-Chief, but in March 1922 an IRA convention confirmed that a majority of IRA units were opposed to the treaty (many of those in favour of it left the IRA and joined the new National Army under Collins). In April 1922 the anti-treaty IRA occupied the Four Courts in Dublin. No action was taken against them until after the general election in June, when the pro-treaty Sinn Féin of Collins and Griffith won a majority of the seats. At the end of the month, using howitzers borrowed from the British military, the National Army opened fire on their former comrades. A bitter and bloody civil war commenced, with IRA units around the country taking sides: the majority around Dublin and in the large urban centres supported Collins; rural units more often were anti-treaty, the most vehement being in the south of the country. Militarily, the pro-treaty side had the upper hand as they were well supplied with armoured vehicles and artillery from the British, weapons that not long before had been used against them. But the pro-treaty forces suffered two huge blows in August 1922, first when Arthur Griffith, then President of Sinn Féin, suffered a brain haemorrhage and died on the twelfth of that month (surprisingly, Griffith merits little more than a mention in most Irish museums) and, more dramatically, ten days later, when Collins was assassinated.

Like Tone, Collins is more than the sum of his parts; he was convivial and popular, ambitious and arrogant. In some respects, he fits Cork County Board's recent definition of 'Corkness', 'that air of confidence just on the right side of arrogance – an unparalleled pride and [an] insatiable desire for Cork to be the best at absolutely everything'. As historian Anne Dolan has observed, 'His past was the stuff of legend, his unlived future a rod to beat mediocrity's back. He was glamour and adventure . . . he was potential unfulfilled and his memory was at the mercy of every man who invoked his name.' I wanted to see how museums and heritage sites dealt with the complex and often controversial commander-in-chief. Would I find a caricature, a larger-than-life romantic figure, more Liam Neeson than Michael Collins, or a nuanced interpretation of his life, death and legacy?

In the National Museum the greatcoat that Michael Collins was wearing on the day he was assassinated is on display, close to the bloodstained shirt James Connolly was wearing when wounded in the GPO. There is something almost relic-like about the preservation of bloodstained clothing, as if physical evidence of blood sacrifice makes them more precious – these are the foundation garments, the holy relics of the Irish state. Despite the fact that Collins was never a prisoner there, Kilmainham Gaol displays a lock of hair taken from his head after he was killed, his cane, the scapular he was wearing when he was shot and his hair brush. In the Military Museum in Collins Barracks in Cork there are letters Collins wrote, his hip flask and the clock from the Leyland touring car he was in when it was ambushed at Béal na mBláth. In death, even the mundane carries so much more meaning.

My own Michael Collins trail around West Cork began in Clonakilty. The street names recall dead patriots, local and national. I doubt there's a greater concentration of republican heroes commemorated in street-name form anywhere in Ireland. There's Wolfe Tone Street, Emmet Square, Rossa Street, Ashe Street, MacCurtain Hill, Clarke Street, Kent Street, Casement Street, Connolly Street and Pearse Street, all within a ten-minute walk of each other. Much later, Collins did get his own bit of road when a section of Western Road was renamed in honour of him, though it quickly fell into disuse. It appears that Clonakilty has had (at best) an ambivalent attitude to Collins, as most of the IRA units in Cork were fervently anti-treaty and, while Collins was a local lad, so too were those who killed him. In 1987 Tom Lyons, the author of a history of Clonakilty GAA, was forced to remove a page that suggested that the young Collins had played football for the club. When, at a club AGM in the early eighties, for 'pure blaggarding', Lyons and some friends suggested that the GAA pitch in Clonakilty should be renamed the 'Michael Collins Memorial Park' there was 'míle murder' and one of the club's officers leapt to his feet, declaring such a thing would happen 'over my dead body'.

It seemed that several generations had to die before the town could countenance recognizing Collins at all. It took until 2002, eighty years after his death, for a statue to be erected, and even then much of the money raised came from the Irish diaspora, not from local pockets. The bronze Collins stands frozen in time as if mid-speech. His posture, energetic and emphatic, recalls the famous film footage of him on the stump outside O'Donovan's Hotel during the general election campaign just weeks before he was assassinated. The statue was unveiled in front of a crowd of ten thousand (twice the population of the town), though I suspect that, while some were there to honour the dead commander-in-chief, many more turned up to get a look at Liam Neeson, who'd come to sprinkle a little bit of Hollywood dust on the occasion. Neeson had portrayed Collins in Neil Jordan's 1996 film, and he played to the gallery, giving a good impression of the 'Big Fella', who, had he been there, would have been a small fella by comparison – Collins was about half a foot shorter than the six-foot-four Neeson. But it is no accident that the wording on the plinth takes pains to include everyone, whatever side they were on: 'This monument is dedicated to the memory of Michael Collins and all those who contributed to the struggle for Irish national self-determination during the Easter Rising, the War of Independence and the Civil War, 1916–1923'. In Clonakilty, even a statue dedicated to the memory of Collins has to be neutral.

Michael Collins lived on Shannon (now Emmet) Square in Clonakilty with his sister, Margaret, and her husband, Patrick O'Driscoll. Popular memory suggested that they lived in the first of a terrace of Georgian houses on the west side of the square, and so sure was Clonakilty Town Council that they put a plaque on the house which read 'Michael Collins lived in this house 1903–1905'. For most people, including Cork County Council, the plaque was proof enough, and in 2016, after spending €4.5 million on the renovation, Michael Collins House opened to the public. But it turns out that memory is fallible and the renaming and renumbering of the square some years after Collins's death had caused confusion. It's now generally accepted that Collins never lived in Michael Collins House – in fact, it's a fair bet that he never set foot in it. The plaque has since been revised to read 'Michael Collins lived in Emmet Square'.

Navigating the museum was occasionally a challenge – at one point I had to kneel on the floor to read a timeline of Ireland between 1798 and 1922 that sits under a scale model of Clonakilty in 1900. And that's part of the problem: the museum isn't quite sure what it wants to be. It seems afraid to engage with Collins in anything more than a cursory fashion and instead tries to link phases of Irish republicanism in West Cork in a narrative that the Young Irelanders would be proud of. Visitors learn about three West Cork men (and their associated women), beginning with Tadhg an Astna O'Donovan, a United Irishman whose pike-wielding statue can be found facing off cappuccino drinkers in Astna Square. O'Donovan fought in the Battle of the Big Cross in 1798, a few miles east of Clonakilty, which was more skirmish than battle. There's also a display about the Fenian Jeremiah O'Donovan Rossa, and his wife, Mary Jane Irwin, who was from Clonakilty. (In fact, an exhibition about Mary Jane Irwin might be the most interesting of them all, for she was a remarkable woman – a committed republican, poet and newspaper editor.) The top floor takes a closer look at Collins's life and death, though so careful is the museum not to attribute blame to anyone for his death that visitors might be forgiven for coming away with the impression that Collins died in a freak accident in which a passing bullet collided with his head. Despite the name, Michael Collins House seems a little embarrassed by itself, afraid of stirring up old emotions. Perhaps, I thought, Clonakilty isn't the best place for a Michael Collins museum.

From Clonakilty I travelled a few miles east to the Michael Collins Centre, a passion project which has been run by the Crowley family (who are distantly related to the Collins family) for the last twenty years. It's an idiosyncratic place. Like so many private sites, it reflects not just the stated focus of the site but also the random interests of the owners. In one glass case on the wall of a repurposed farm building there is a signed photo of Charles Stewart Parnell, the Leader of the Irish Parliamentary Party, a German medal minted to celebrate the sinking of the *Lusitania* in 1915, a shillelagh, a German belt buckle from the First World War and a stick for shining army buttons. On the walls are reproductions of newspapers about Michael Collins; above the fireplace a large, hand-painted map of West Cork shows 'Michael Collins Country'.

The visit began with the screening of a short film which showed Tim Crowley, a genial, grey-haired man, standing at all the main West Cork sites associated with Collins – Woodfield, Sam's Cross, the All Inn, Clonakilty, Béal na Bláth. Then Tim appeared in real life and began what he assured us would be a thirty-minute lecture accompanied by slides (though we were warned that he could happily talk about Collins for several hours). The Collins that Crowley described was much more real than the cardboard version I'd met in Clonakilty. Crowley's Collins was vibrant, passionate and quite possibly drunk when he died. Tim Crowley wasn't afraid to speculate about who killed him, or discuss the messy details of the Civil War, though he also told the moving story of Jim Hurley, a member of the ambush party who assassinated Collins. Soon after the Civil War ended, Hurley visited Collins's brother Seán to seek forgiveness for his role in the assassination. The two men stayed in contact and, when Seán Collins died in 1965, Hurley, who had terminal cancer, requested that he be buried beside him, which he was, just one week later.

Beyond the old farm buildings which house the exhibition there is a grassy track, and just around a bend in the track there is a reproduction of the Béal na Bláth ambush site, complete with replicas of Collins's convoy of a Crossley Tender, Leyland touring car and Rolls-Royce armoured car. On 22 August, Collins and his convoy had travelled from Cork city as far west as Skibbereen, meeting supporters and family members before heading back to the city. A barricade had been set up on the road at Béal na Bláth by anti-treaty IRA men and an ambush party of about thirty waited there, hoping that Collins would return that way. As night began to fall there was still no sign of the returning convoy and the ambush party headed home, leaving a handful of men behind to remove the barricade. What followed has been the subject of speculation for almost a century. All we know for sure is that when Collins's convoy arrived, the two sides exchanged gunfire for about half an hour and there was one casualty – Michael Collins himself, killed by a bullet which hit him in the head. There was no need for Collins's convoy to stop and fight that evening; by the time it arrived, the barricade had been removed and the convoy could have blasted its way through. Emmet Dalton, Collins's aide-de-camp, recalled that when the first shots were fired he told the driver to 'Drive like hell!', but that Collins overruled him, insisting that they stop and fight, a decision that, Professor Joe Lee believed, showed the commander-in-chief 'behaving more like a cowboy than a head of government', and possibly a drunk one at that. It is now generally agreed by historians that the fatal shot was fired by Sonny O'Neill (a former member of the RIC who had fought with the British Army during the First World War, had joined the IRA in December 1918 and chose the anti-treaty side in the Civil War). But such academic agreement hasn't filtered out more broadly. In one day in West Cork I was told several different versions of Collins's assassination, each related with great verve and conviction. Everyone with an opinion on who shot Collins was convinced that they are right. A few stood by the old (and certainly untrue) story that Collins's death had been ordered by de Valera. Far more agreed that O'Neill fired the fatal shot, while others placed the blame squarely on Collins's own men – some claiming that he was killed by 'friendly fire', others even asserting that he was deliberately shot by Emmet Dalton.

From Michael Collins Centre I drove thirty kilometres north to see the real Béal na Bláth. The landscape there has been altered so extensively that the reproduction at the Michael Collins Centre offers a more realistic version of the site as it looked in 1922. At Béal na Bláth the bend has been straightened, the road widened and space created to erect a memorial to Collins. The memorial is unprepossessing – a few steps lead up to a cross, surrounded by black iron railings topped with gold finials. Collins's name and date of death are inscribed in Irish around the base. Strewn at the foot of the cross were a selection of rosary beads, holy water in a plastic Virgin Mary, candles and some small tricolours. When I passed that way some weeks earlier, there was a different batch of votives. I often wonder, not so much about the people who leave these signs of devotion, but about those who take them away. Who are they and what do they do with the items they remove?

Given the ambivalence I found in West Cork, it's not a huge surprise that Collins isn't buried there. His body was put on board the steamer *Classic* in Cork and sailed up the coast to Dublin. For three days he lay in state in Dublin's City Hall while thousands filed past the open coffin. On the day of his burial more than three hundred thousand people lined the streets as his body was taken from the pro-cathedral on Marlborough Street to Glasnevin Cemetery. His grave, as every tour guide will tell you, is the most visited in the cemetery. Collins was laid to rest in the Army Burial Plot without any marker to publicly indicate the precise location. In 1935, Seán Collins requested permission to erect a headstone and, following four years of heel-dragging by the Dublin Cemeteries Committee, it was finally granted. Erected in 1939, seventeen years after his death, Collins's gravestone is an unremarkable twelve-foot-tall plain limestone cross. The simplicity of the memorial may reflect the wishes of the Collins family, but they were restricted in what they could do as Éamon de Valera (then Taoiseach) forbade the erection of any ostentatious memorial. The grave stands alone, a little away from all the other plots, though now rather too close to the toilets and coffee shop. There are always fresh flowers on it and, famously, a French woman, inspired by the film *Michael Collins*, makes an annual pilgrimage to lay flowers on Collins's grave.

Collins remains an elusive figure. Despite trailing him around the country from birth to death, I don't feel I have any sense of who he really was. For some, he's a sell-out; for others, he's the Hollywood hero; for members of Fine Gael, he's their founding father (despite the fact that he died long before the party was founded); still others regard him as a pragmatic politician. At the military museum in Collins's Barracks in Cork, the curator, Sergeant Denis McGarry, described him as 'the hijacked hero'. Hero or not, his memory has certainly been hijacked by many to suit their own ends. What I found was not a caricature, but rather a chameleon who could transform from soldier to spymaster to bon vivant to statesman with ease. In some ways, Wolfe Tone is much more vibrant, in part because he left behind many more words. While his voice can be heard through his extensive diaries, memoirs and political works, we hear only snatches of the real Collins in letters, speeches and the recollections of others.

Many museums only have space to display a tiny fraction of their collection at any one time. Thousands of objects are in storage, hidden in boxes and cupboards, awaiting a turn when fashion or interest dictates. On my journey to Ireland's dark side I occasionally got to have a sneak peek behind the museum display. I had one of those rare treats when my friend Brenda Malone, curator of military history at the National Museum, allowed me into the stores – a series of rooms lined with grey cupboards full of thick cardboard archive boxes containing thousands of artefacts, all carefully wrapped in acid-free tissue paper. There's a particular whoosh sound, like air escaping from a vacuum, that's heard every time an archive box is opened. I like to think of it as a little bit of the past escaping into the present.

I was, unsurprisingly, particularly interested in the 'Death and Human Remains Cabinet', though a little disappointed to find that it's not full of skeletons suspended on hangers like unused suits. Rather, it's a collection of objects associated with death, or more correctly, objects associated with people who have died. I was there to see two particular items – the death masks of Theobald Wolfe Tone and Michael Collins. This was my chance to bring them face to face – literally. They bookend the Irish revolutionary story. Brenda carefully unpacked the death masks of Tone and Collins and placed them in front of me. I'd seen a death mask of Tone from a distance in St Michan's but, up close, he looks gaunt, his face, small and narrow, dominated by a long, prominent nose overshadowing thin lips. By contrast, Collins has a larger, fuller face, his hair swept back off his forehead by a gauze bandage. Sitting in a museum store was as close as I could get to these two hugely influential men. It was a strangely moving experience, for it felt that the masks did contain something of the essence of the men. Despite their plaster-cast quality, they somehow feel more real, more alive, than any painting or photo.

Despite the century that divides them, there is much that unites Tone and Collins, both in the way they presented themselves in life and in the way they've been recalled in death. They represent the fight for Irish freedom, though the Ireland that emerged from the War of Independence was not one either of them envisaged. In person there are similarities too: both saw themselves as soldiers and, despite neither having any real training in either strategy or fighting, they ended up in military uniform – one an adjutant-general in the French Army, the other Commander-in-Chief of the National Army. Alongside their military, political and patriotic commitments, they found time for carousing and dalliances, which has certainly added to their appeal. Both were flawed, which perhaps makes them more relatable, more human, than many of the other more sober, staid figures who fill the pantheon of Irish republican heroes. And both died young – a vital factor in becoming a national hero – and in controversial circumstances. For many years, Irish nationalists and republicans refused to accept the fact that Tone had died by suicide, while debate still rumbles over who fired the shot that killed Collins. Tone was Protestant, something that didn't suit a Catholic nationalist version of what it was to be an Irish republican – and so his religion was often glossed over. Tone and Collins: their complexity, their contradictions, their vision, their youth and their death combine to make them ideal totems for Ireland's dark narrative.

Those that I fight I do not hate

W. B. YEATS,
An Irish Airman Foresees His Death

In 1996 my mother's friend Barbara had the unhappy task of clearing out her family home following the death of her mother. One afternoon, as she sifted through the piles of papers, boxes and bags she'd taken down from the attic, she opened a wooden box containing documents and letters relating to a man named Thomas Walker. Though they shared a surname, Barbara had no idea who he was. After some digging around she discovered that he was her uncle, an uncle she had never heard of, who had joined the British Army and fought in France in the First World War.

Thomas Walker was born in Dublin on 21 February 1895. He grew up with his parents and nine siblings in 38 John Dillon Street, a four-room end-of-terrace house close to St Patrick's Cathedral. By 1911, aged sixteen, he was working as a silk weaver. Four years later he joined the Royal Dublin Fusiliers and later transferred to the Machine Gun Corps. He was dispatched to France in early March 1916. On 30 April, while he was on guard duty, German soldiers released a cloud of chlorine and phosgene gas. Eighty-nine men died and Thomas was one of more than five hundred hospitalized. In a letter to his family from hospital, he described it as a 'cowardly' attack, but his primary concern was to reassure his parents that he was regularly saying the rosary and to enquire after his beloved Olympia Football Club – a club based in the Liberties in Dublin. He hoped that they'd won the Leinster Senior Cup (they had not).

Thomas was one of over two hundred thousand Irishmen who enlisted in the British Army and fought in the First World War. They joined for a myriad reasons, some for travel and adventure, others because of politics or religion, others still because of opportunity or friendship. Many Irish nationalists joined in re-sponse to a speech made by John Redmond, the Leader of the Irish Parliamentary Party, to an assembly of Irish Volunteers at Woodenbridge, County Wicklow, in September 1914, in which he encouraged members to go 'as far as the firing line extends' because the war was 'undertaken in the defence of the highest principles of religion and morality and right'. He was certain that when the war ended Ireland would get Home Rule. On the other side, unionist members of the Ulster Volunteer Force (a paramili-tary force founded in January 1913) joined the 36th Ulster Brigade, equally convinced that their loyalty to the Empire would prevent Home Rule from taking place. While Thomas was on the Front in France, his younger brother Michael, a member of the Irish Volunteers' Dublin Brigade, took part in the Easter Rising. As Thomas recovered from the gas attack at a convalescent camp in Boulogne, the execution of the rebel leaders in Dublin began, and everything changed. In the wake of the executions, the mood in Ireland turned decisively against the British, leaving thousands of Irishmen fighting in the British Army in an impossible position. The political landscape at home had been transformed and it was no longer acceptable for Irish nationalists to wear the uniform of the British Crown. A general silence descended and there was a bitter irony in the fact that many had fought and many had given their lives, thinking it would advance the case for Irish freedom, and now their reward was to be written out of history.

For decades after the end of the First World War Ireland had an ambivalent relationship with the Irishmen who had fought as members of the British Army. In the mid-twenties a proposal to erect a war memorial in Merrion Square in Dublin was dismissed by the Minister for Justice Kevin O'Higgins (who himself had a brother who died in the war). O'Higgins told the Dáil that: 'No one denies the sacrifice, and no one denies the patriotic motives which induced the vast majority of those men to join the British Army to take part in the Great War, and yet it is not on their sacrifice that this state is based and I have no desire to see it suggested that it is.' And so a site in Islandbridge, much further away from the Irish parliament, was settled upon. The War Memorial Garden was designed by the architect Edwin Lutyens, who was responsible for the design of over 175 war memorials and cemeteries. It was completed in 1939 and dedicated 'to the memory of the 49,400 Irishmen who gave their lives in the Great War, 1914–18'. The memorial was largely ignored by the state and, although some Remembrance events were held there, the site was neglected. In the fifties, Irish Republicans twice unsuccessfully attempted to blow up the 'Cross of Sacrifice' and over the decades Lutyens's memorial became dilapidated and the gardens overgrown and full of rubbish dumped there by Dublin Corporation. It wasn't until the mid-eighties that work was undertaken to restore it; the memorial garden reopened in 1988. In 2006, on the ninetieth anniversary of the Battle of the Somme, President Mary McAleese laid a wreath at the memorial and in 2011 both McAleese and Queen Elizabeth II attended a ceremony in the gardens.

The building of the memorial outside the city centre and the fact that it was neglected by the state for half a century highlight the marginalization of the First World War in Irish official and popular memory. There was little public or private appetite for telling tales of a war fought on a foreign field, but just as the War Memorial Garden has received a new lease of life, now museum and heritage sites regularly incorporate stories of the Great War into their exhibitions. There are thousands of stories to be told about the reasons men joined the army, their accounts of war and their experiences of returning home to a country that refused to welcome them, but it's only in the last couple of decades that the threads of these stories are being discovered and slowly woven into the broader fabric of the history of Ireland in the early twentieth century.

Quite apart from the complex politics associated with the First World War, the Easter Rising and the division of Ireland into two jurisdictions in 1920, one reason why Irish involvement in the war did not get much attention is because none of the action took place on Irish soil, although there was some preparatory activity.

During the war Cork Harbour was a significant operations base for the British Navy and in 1917 a flotilla of American destroyers joined the British ships stationed there. And there are scars on the landscape that bear witness to the exercises which soldiers took part in before being sent to the Front. Training trenches were dug across the country, and evidence of these has been found on Spike Island, at Kilworth, County Cork, at the site of the Crinkill Barracks near Birr, County Offaly, and at the former army camp at Ballykinler in County Down.

By the beginning of July 1916, Thomas Walker had recovered from the gas attack and was preparing to take part in one of the largest battles of the First World War – the Battle of the Somme. There is a replica of one of the Somme trenches at Cavan County Museum and one day in March I borrowed Lucy from school and we went to visit. The museum is in a former convent and, before venturing into the trenches, we looked at the small First World War exhibition, housed in a few rooms that were once nuns' cells. One room tells the story of the Moore family: four of the five children took part in the war. One of them, Charles, was killed. Medals, letters and postcards sent home by the Moore siblings are on display, including a card brightly embroidered with flowers and the words 'Erin go Bragh' ('Ireland Forever'), possibly indicating a political allegiance. Also on display is a document signed by Arthur Kenlis Maxwell, Lord Farnham of the Farnham Estate in Cavan, when he was a prisoner at Karlsruhe in Germany, in which he gives his word that if he went on a walk outside the prisoner-of-war camp, he would not try to escape.

From the exhibition, we headed outside to the trenches. It might seem incongruous to have a replica trench on the outskirts of Ballyjamesduff, but as the museum makes clear there was a close Cavan connection to the Battle of the Somme. Major-General Oliver Nugent of Mountnugent, County Cavan, commanded the 36th Ulster Division, and at least twenty-seven Cavan men were killed on the first day of the battle alone. The narrow trench, a replica of the one used by the Royal Irish Fusiliers in July 1916, is over 350 metres long, with wooden posts supporting layers of sandbags and mounds of earth. It caught Lucy's imagination far more than any artefact in an exhibition case. She climbed on to the firing step, peered through a periscope on to the barbed-wire-strewn no-man's-land and took aim with a rifle at the far-distant enemy. We explored the 'kitchen' area with its rusted tins of food and the makeshift first-aid camp. Lucy was horrified to discover that biscuit tins were used as toilets, and when she read about the rats, mice and lice that infested the trenches she broke into a run. 'Hurry up!' she shouted from the safety of the exit, 'I think they might actually have rats and lice in there!'

As I walked through the trench I thought of Thomas, who would have been in a similar trench in early July 1916. Visiting the trench gives a sense of what life on the Front might have been like, but it can't convey the horror, fear and brutality of the real experience (and I'm grateful that it can't). On 13 July 1916, two weeks into the battle, Thomas Walker was fatally wounded. He was twenty-one years old.

Like many other families, in the years that followed, the

Walkers preferred to grieve in private. Perhaps it was easier to stay silent, and Thomas's name was never mentioned. His niece Barbara found his letters home, his rosary beads, medals and pipe in the long wooden box alongside the family Bible and various birth and death certificates, where they had been safely and quietly stored eighty years earlier. In 1999 my parents accompanied Barbara on a very personal pilgrimage to Puchevillers British Cemetery, about twenty kilometres north-east of Amiens, where Thomas Walker is buried alongside 1,700 other British Army soldiers. The cemetery, also designed by Lutyens, is hidden away, down a narrow country lane just outside the village. Five large whitebeam trees near the 'Cross of Sacrifice' provide shade, while rambling roses and flowering shrubs line the perimeter walls. Thomas's gravestone is like thousands of others – white Portland stone with the regimental badge of the Machine Gun Corps at the top and a cross carved in the centre. It had remained unvisited for eighty-three years.

In a diary entry, written on his first day in France in March 1916, Thomas had written: 'When far away and friends are few/ Remember still I love you.' Before leaving for France, Barbara had had those words engraved on a small plaque below Thomas's name. She brought the plaque with her to Puchevillers and laid it on his grave.

Maritime
Disasters

࿇

'Hear the sound of the sea,
like mourning beside a grave'
ATTRIB. ST COLMCILLE

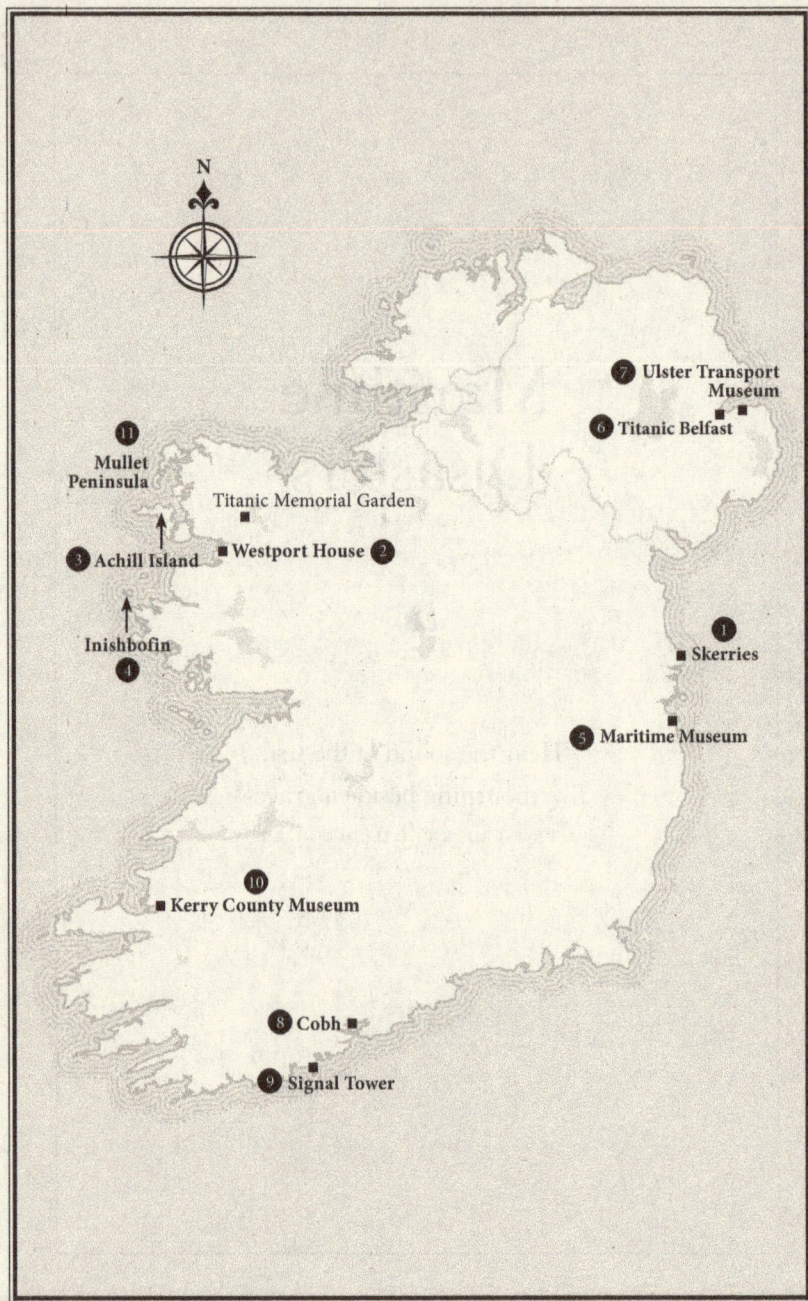

N

7 Ulster Transport Museum

6 Titanic Belfast

11 Mullet Peninsula

Titanic Memorial Garden

3 Achill Island   Westport House **2**

Inishbofin

4

1

Skerries

5 Maritime Museum

10

Kerry County Museum

8 Cobh

9 Signal Tower

I grew up beside the sea in Skerries and, as a child, every day I ran along the beach, passing Shenick, Colt and St Patrick's islands on my way to my friend Gráinne's house. There's a Martello tower on Shenick Island and at low tide we'd walk out to it and, using an old frayed rope, haul ourselves inside. There, inside the dilapidated tower, we'd terrify each other by telling tales of the smugglers and pirates who used nearby Smugglers' Cave to stash their stolen treasure. Sometimes, after we had waded back from the island, we would scramble along the rocky seashore towards Loughshinny to the cave and dare each other to go inside. From the outside, it's pretty unprepossessing, a small gash in the low cliff. I was never brave enough to go much further than a few yards in, as the passageway gets very narrow and you have to crawl along it before it opens up into a wider tunnel. Rumour had it that a green serpent guarded a huge pile of glittering treasure in the large cavern at the end of this tunnel – gold left there by the handsome pirate 'Jack the Bachelor', and jewels and bags of silver coins from nearby Baldungan

Castle, hidden when the castle was attacked in the 1650s. All that remains of the castle today are the ruins of the church; all else was destroyed by Cromwell's men and time. I knew the story about the serpent couldn't be true, for hadn't St Patrick got rid of all the snakes and serpents, but I believed that a tunnel linked the castle to the cave and that all the treasure was there, waiting for us to discover it. I believed it, but I wasn't prepared to investigate. Other, braver friends hunkered down and crawled along the passageway with torches clasped in their teeth. They returned with stories about how the cave was huge, and there really was a tunnel carved through the rock by human hands. There had definitely been pirates there and, if they had time, and if we didn't all have to be home for our tea, they'd have kept exploring until they'd found the treasure.

'Jack the Bachelor' is one of two pirates I knew about as a child. The other is Grace O'Malley, the 'pirate queen', whose real life is overshadowed by myth and legend. While I was certain that 'Jack the Bachelor' was real – I once cycled to Kenure Cemetery in Rush to see where he is buried (close to the memorial to the cholera victims of the 1840s) – for years I thought Grace O'Malley (also known as Gráinne Mhaol) was from a storybook. Tales about her tumbled together with those of Queen Maeve, the Children of Lir, Cúchullain and Tír na nÓg. I know of her from cartoon images and fantastical stories, not from any history book, but (fictional or real) she caught my imagination, for in my childish version she proved to doubting men that women could do anything as well as (and probably better than) they could. She robbed from the rich to give to the poor. She was feisty, smart and successful, and seemed a pretty good role model to me. Not all her contemporaries were so impressed, with Lord Drury, President of Munster, acknowledging that she was 'famous for her stoutness of courage' but also that she was 'chief commander . . . of thieves and murderers at sea'. The Queen's Privy Council agreed and noted in 1578 that she was a 'notorious offender', though in the stories I recall from childhood she became friends with Queen Elizabeth I. I was interested to see how museums portrayed her now. Would she be cast as a romantic mythic creature or be afforded the same respect as a male contemporary?

The late sixteenth century was a turbulent time for Ireland, on land and on sea. Ireland's most famous pirate, O'Malley, was born around 1530 and came from a seafaring family whose wealth and status was derived largely from trade, piracy and taxing any boats that sailed through waters dominated by their ships. But their power wasn't confined to the sea; the family controlled much of the land around Clew Bay, including Achill and Clare islands, and owned a cluster of tower houses – these were pirates with a lot of land. Yet it's surprisingly hard to learn anything about Grace O'Malley. I started in Westport House, County Mayo, a Georgian pile nestled in a crook of the Carrowbeg River. It's a much-amended country house in pleasant grounds not far from the centre of town (the village of Cathair na Mart had once been located directly outside the house, but in the mid-eighteenth century the owner, John Browne, decided he'd rather have fine gardens and views outside his house, and so the village was swept away and the new town of Westport was built a mile inland). It seemed a little strange to search for tales of the pirate queen not on the coast or in a castle but in a 'big house', but the house itself is built on the site of one of the O'Malley castles, and traces of it remain in the basement. The house is a genial mix of clutter, bad paintings (some attributed to famous painters who ought to have tried harder), draughty bedrooms, fusty Edwardian furnishings, waxworks of famous visitors, chinoiserie, Wedgwood and silverware, and a staircase in Sicilian marble is dominated at the turn by a giant 'Angel of Welcome', but there's no immediate sign of drama on the high seas. You have to go below stairs for that.

Downstairs, alongside the kitchen and the café, is the 'Grace O'Malley Dungeon'. Most of it is inaccessible, but there's a brightly costumed dummy at one end, illuminated by dramatic lighting. There's no explanation of who the prisoner is supposed to be, but presumably it's not Grace herself, for she's unlikely to have spent any time imprisoned in her own dungeon. Along the walls of the café there is a Grace O'Malley 'exhibition' consisting of half a dozen information panels, where I learned little more than I already knew from my childhood tales – that because her father had refused to allow her to board his ships for fear her long red hair would get caught in the rigging, she shaved it off, an act reflected in her nickname, Gráinne Mhaol (Gráinne the Bald). That aside, the dungeon is largely an opportunity to feel cold and damp, which, given that we are in Mayo in February, can be done anywhere for free. Elsewhere, the O'Malley connection is shored up by the presence of a large statue of her and a family tree demonstrating that her descendants – the Browne family – eventually built Westport House.

Westport House exemplifies contemporary attitudes to O'Malley, who makes at most a fleeting appearance in Irish museums. In part this is no surprise, for there are huge challenges in telling her story (or that of any pirate). Pirates did not tend to keep many written records, and a life spent at sea left little material evidence. You might as well try to build a museum for selkies. But it's a shame, as O'Malley was a significant figure in Gaelic Ireland from the late 1540s until her death in the early seventeenth century. Over the course of more than half a century she withstood repeated attacks both from Gaelic chieftains and English representatives in Ireland. Her greatest challenge came from Sir Richard Bingham, Queen Elizabeth's representative in Connaught. Bingham was determined to destroy O'Malley by confiscating her land, arresting her youngest son and threatening to have him tried for treason. She took the dramatic step of appealing directly to the Queen, and in 1593 O'Malley sailed from Clew Bay to London. The two women met in Greenwich and terms were agreed that were favourable to the Irishwoman – her son was released and her lands returned. Queen Elizabeth's generosity may have been partly because she grudgingly admired O'Malley – another woman who had succeeded in a world designed for men – but it's more likely to have been pragmatic; she didn't see her as a long-term threat, and limited concessions were a small price to pay for potential peace in Connaught. In a letter to Bingham, Queen Elizabeth referred to O'Malley as an 'aged woman' and declared that she was convinced that she would 'as long as she lives, continue to be a dutiful subject', the implication being that she might not live long, though in the end O'Malley, like Elizabeth herself, lived on until 1603.

I find stories of drownings very emotive, particularly those associated with fishing, perhaps because when I was a child there was still a vibrant fishing fleet in Skerries Harbour and a fish-processing plant in the town. Many of my classmates had fathers, uncles and brothers who were fishermen, and every few years some disaster would strike, with boats and crew lost at sea. A memorial at the harbour, beside the lifeboat station, recalls over 270 people who have drowned off the coast since the mid-nineteenth century. In this respect, it's a typical Irish coastal town. A ring of sorrow encircles Ireland, binding together coastal communities who have suffered maritime tragedies like beads on a rosary. Every inch of the coast has its harrowing seafaring stories to tell, every community bears invisible scars. On Achill, near Dogoort, on the crest of the hill just beyond the Silver Strand, an information board recounts the catastrophe which befell the island in June 1894. That summer, hundreds left Achill, as they did every year, to pick potatoes in Scotland. The islanders were taken by hooker to Westport to board the SS *Elm*, a Glasgow-bound steamer. As one of the hookers approached the *Elm* she listed heavily, turned on her side and tipped the passengers into the sea. Thirty bodies were pulled from the waters of Clew Bay. Two were never recovered. At the same time, there was good news coming to Achill. The train line to Achill Sound (linking the very west of the country to the rest) was just about to open, and on 16 June 1894 the first train duly arrived in the station. It should have carried jubilant passengers. Instead it bore the bodies of the drowned. The grief-laden train journey fulfilled the Mayo seer Brian Rua Ó Ceabhain's seventeenth-century prophecy, which foretold that 'iron wheels with smoke and fire' would carry home the dead. In fact, the train line to Achill was bookended by tragedy. While the first train returned the victims of the Clew Bay disaster, the last train, in 1937, brought back the bodies of ten Achill men who had died in a fire in Scotland.

The bodies of those who drowned in Clew Bay are buried in one grave in Kildavnet Cemetery, beside a castle that once belonged to Grace O'Malley. The cemetery slopes gently down to the sea, and my visit disturbed several sheep quietly grazing on the grass verges. Twenty-five of the victims were women. One can only imagine the hardship these women endured if they were prepared to move to Scotland for several months to undertake hard, physical labour, for low pay and in poor living conditions, but there's something particularly moving about Mrs Doogan and Mrs Mulloy – the only two married women listed. It seems conditions on Achill were so desperate that even married women left their families for months on end in order to earn enough money to feed and clothe them.

One of the saddest seafaring tragedies occurred in October 1927, when, in the words of the poet Richard Murphy:

> A storm began to march, the shrill wind piping
> And thunder exploding, while the lightning flaked
> In willow cascades, and the bayonets of hail
> Flashed over craters and hillocks of water.

The unexpected storm wrecked a number of currachs and hookers and drowned forty-four men from Rossadilisk, Lacken Island and Inishbofin, Aran and the Inishkea islands. The 'Cleggan Disaster' was devastating, in part because of the number of lives and boats lost, but also because those losses shattered vulnerable communities.

At the pier in Cleggan on a sunny summer afternoon I boarded the ferry, *Island Discovery*, with Al, Jessie and her children and her father, Jerry. The ferry, which takes day-trippers and residents to Inishbofin, is about three times the length of the currachs, and twice the length of the hookers that were caught in the storm in 1927. We sat outside on the deck while waves splashed enthusiastically against the hull, but on such a calm, clear day there was no sense that the sea could turn and swallow you whole – many of the forty-four who drowned that October night were never found. As we approached the harbour we could see, on our right, the picturesque ruins of the fort built in the 1650s; it was briefly used as a prison to hold Catholic clergy before they were transported to Barbados by the Cromwellian regime. The story of the Cleggan Disaster is told in a small display housed in a former store near the pier, and there's a memorial at East End beach to the nine men who set sail from there that night and never returned. A local dog scampered at our feet as we walked along the beach with its pristine white sand, collecting tiny cowrie shells as the water gently lapped against our feet. Gazing at the tranquil ocean, it was impossible to imagine the desolation that must have been felt by the island communities who watched as the 'bayonets of hail' destroyed them.

The storm damaged the community on the Inishkea islands off the coast of Mayo even more than those on Inisbofin. The two Inishkea islands had a combined population of under two hundred and fifty in 1927, and of the twelve men who set out in six currachs from those islands, only two returned home alive. The bodies of those recovered from the sea are buried in one grave in Fallmore Cemetery on the Mullet Peninsula. Families and extended families were shattered – of the ten men drowned, there were four Ó Muineacháins and three Ó Raghallaighs. This loss devastated the community and precipitated the forced evacuation of the island. In 1911, there were 24,700 people living on 124 islands off the coast of Ireland, but by 2016 that figure had dwindled to 8,756 residents on 64 islands. The Inishkea evacuation of 1934 was the first of many encouraged or forced evacuations of islands along the west coast. Over the next few decades, others followed, including Rutland, Inishmurray, Inishark and the Blaskets.

The two most famous maritime disasters in Irish history are undoubtedly the sinking of the *Titanic* in 1912 and the *Lusitania* in 1915, but there's another, now largely forgotten, though at the time it received enormous attention. In October 1918, the mail-boat HMS *Leinster* was attacked and sunk by a German U-boat while sailing between Dún Laoghaire and Liverpool. Of the 771 passengers and crew on board, over 500 were lost. The story of the *Leinster* is told in the former Mariners' Church in Dún Laoghaire, now a maritime museum. A mix of panels, artefacts (including a spittoon from the ship) and screens tell a story that's broadly sympathetic to everyone, including the U-boat crew, and sets the sinking in the context of the war and German attempts to erect a maritime blockade of Britain and Ireland. (By the time of the First World War, it's not castles that are besieged and starved, but whole countries.) Despite the fact that the sinking made headlines around the world, it has largely been forgotten, perhaps in part because 350 of those who drowned were Irish men serving in the British Army. In the turbulent years of the War of Independence and the Civil War that followed, commemorating the loss of British military personnel was hardly a priority for the fledgling Irish state and the tragedy faded from view as the Free State and later the Republic chose to mark more obvious nationalist anniversaries.

Any exploration of the *Titanic* story has to start in Belfast, the birthplace of the ship. It's fair to say that the city hasn't always got good press. For decades, it was a very hard sell from a tourism perspective. But all that's changed, and Belfast's tourist scene thrives on dark tourism, whether that's black-taxi tours around the murals, a trip to Crumlin Road Gaol or a visit to Titanic Belfast. The city has embraced the story of the *Titanic* with gusto. Here, *Titanic* is not a byword for hubris and disaster but a somewhat perverse source of pride.

In October 2010, I went with some friends to see the building site where Titanic Belfast was being constructed. Only in Belfast, one of them drily pointed out, would they build a monument to the most famous ship in the world in the shape of the iceberg that sank it. In fact, that's a little unfair. The design is intended to combine aspects of the White Star Line logo with the shape of a ship's hull and the crystal structures of an iceberg. The completed building, which cost over £100 million, is majestic in many ways; it rises 130 feet high, all sharp angles and bright reflections. The pools that encircle Titanic Belfast catch the light from the silver aluminium sheets that clad the building and bounce it back up. At the right angle and in the right light, this makes the building look like a ship scything through the waves.

Titanic Belfast bustles like a transport hub. On the day I visited, it was heaving with people of all ages and there were queues everywhere – for tickets, for tea, for toilets. I nosed around the shop looking for something to add to my cabinet of curiosity – a *Titanic*-themed thimble or a box of *Titanic* fudge? I treated myself to both. The shop staff were irrepressibly cheerful. Every customer was greeted with a smile and an enquiry as to how their day had been and how long they were in Belfast for. This gave me an excellent opportunity to indulge my passion for eavesdropping – and my (unscientific) survey revealed that most people buying *Titanic*-themed T-shirts, snow globes, pencil cases and hip flasks were in the city for a handful of hours before re-boarding one of the two cruise ships that could be seen from the *Titanic* slipway. I thought it rather brave of them to spend their few hours ashore learning about the disaster that befell another cruise ship.

Before visiting the exhibition, I explored the slipways where the *Titanic* and her sister ship, the *Olympic*, were built. The attention to detail is impressive. Glass strips mark the outlines of both ships, with benches placed where seating would have been on the deck of the *Titanic*. I couldn't help thinking of the body outline at a crime scene. In May 1911, a hundred thousand people surrounded the slipway to see *Titanic* being launched into Belfast Lough. At the water's edge, I watched a young couple re-enacting the 'I'm flying' scene from James Cameron's film *Titanic* – the romantic scene where Rose stands on the railings at the bow with Jack behind her while Celine Dion opines (contrary to all medical opinion) that 'the heart does go on'. Surely they knew how this love story ends.

Back inside Titanic Belfast, I dodged the enthusiastic staff photographer who tried to persuade me to pose as an excited passenger about to embark on the voyage of a lifetime and went back in time to 'boomtown Belfast'. The city was a shipbuilding titan, and the most famous of its shipbuilding companies was Harland and Wolff, which constructed some of the best-known ships in the world for the White Star Line, including HMS *Belfast* and the *Olympic*. The busy exhibition galleries replicate something of the hustle and bustle of the city in its heyday. Panels, videos and recordings explore and explain industrial Belfast (though the sectarian employment policies of many of its companies, including Harland and Wolff, are entirely glossed over). It's all beautifully packaged, but, with only a handful of artefacts on display, I was a bit disappointed. Surely there are plenty that could have been used to illustrate the stories? From the Belfast galleries I climbed aboard a slow fairground ride and was transported through a reconstruction of the shipyard, complete with a soundscape of rivets being hammered and steel being bent, before being disgorged to explore the lavish interiors of the great ocean liner. Visitors can step, virtually, into the ship and be taken through it as if in an elevator – from the engine room up through the dining rooms, the lavish first-class cabins and the library to the bridge. Unsurprisingly, the tone changes as visitors move into a gallery that replays stark Morse code messages calling for help when the *Titanic* struck the iceberg. From there, the focus is on the sinking, the investigations into the cause of the tragedy and the public fascination with the story. Some of the myths about the ship are tackled too – it was never advertised as being unsinkable, the rivets were not faulty, the 'Heart of the Ocean' necklace made famous in the film *Titanic* was a figment of director James Cameron's imagination, though there were many jewels on board – one body was found with seventeen diamonds in their pockets. The sense of a journey, both through the construction of the boat and across the ocean, is clear, though sometimes I felt swept along and past lots of the information, for the crowds were a little overwhelming.

The *Titanic* attracted an enormous amount of attention before it ever set sail, and the news of its sinking made headlines across the world. Thirty thousand people gathered to stare at the survivors on board the *Carpathia* when she docked at New York, while police in Halifax, Nova Scotia, were tasked with guarding the bodies of the dead to prevent souvenir hunters stealing their belongings. From the beginning, there was a seemingly insatiable desire for stories, mementoes and trinkets associated with the disaster. Within weeks, commemorative songs, books, postcards, films and handkerchiefs were for sale in Britain, Ireland

and North America. By the end of 1912, over a hundred pieces of music about the *Titanic* had been copyrighted in the United States alone, five books had been published and one film – *Saved from the Titanic* – released. Both Titanic Belfast and a second exhibition, *Titanica* – which is at the Ulster Transport Museum a few miles outside Belfast – display sheet music for songs such as 'My Sweetheart Went Down with the Ship', postcards of the ship printed with the lyrics of 'Nearer My God to Thee' – the tune allegedly played by the band as the ship sank – and posters and extracts from films, including the 1953 epic, *A Night to Remember*, James Cameron's saccharine blockbuster, and *Titanic*, a Nazi propaganda film made in 1943 which portrayed the British as greedy, prioritizing profit over people, and a German officer as the hero who rescued third-class passengers.

At the Ulster Transport Museum, the *Titanica* exhibition is tucked quietly behind the cavernous hangars filled with trams, trains, cars, bikes and planes. A boy of about four ran past me on his way to climb into one of the enormous steam engines, shouting back to his mother that this was the 'best day ever'. While I had had to jostle my way around Titanic Belfast, in Cultra I almost had the place to myself. That's a shame, because *Titanica* has an understated treasure trove of hundreds of artefacts associated with the ship on display. Here, in an exhibition case housing items salvaged from the wreck, are my favourite artefacts from the ship. They are two tiny perfume vials, displayed alongside the business card of Adolphe Saalfeld, a German perfumer who was travelling in first class. In a leather satchel he carried a set of perfume samples which he hoped would help him break into the lucrative American market. Saal-feld survived the sinking, but he never opened an American store. He returned to his wife in England and died in London in 1926. In 2000, Saalfeld's bag was found during a salvage oper-ation, and in 2009 a perfume, Night Star, was created (by a now-defunct company called Scents of Time) based on the sam-ples discovered in the satchel. The story associated with the vials of perfume represents to me all those lost dreams, all the poten-tial wiped out that night in April 1912. For one brief moment, as I contemplated the perfume bottles, it occurred to me that it would have been an even better story if Saalfeld hadn't survived. Clearly, I'd been contemplating misery for too long – perhaps the woman in Dublin Castle had been correct and too much exposure to misery did indeed have a detrimental effect.

From the birthplace of the *Titanic* I headed south to Cobh, in County Cork, then called Queenstown. It was the *Titanic*'s final stop before heading for New York. Like in Belfast, there are two exhibitions – the Titanic Experience, housed in the old White Star Ticket Office, and another across the road at the Cobh Heri-tage Centre. On 11 April 1912, the final 123 passengers boarded the tenders which took them out to the ship, the vast majority of them third-class passengers. A lucky eight disembarked – seven first-class passengers and John Coffey from Cobh. Coffey was a stoker on the *Titanic* and it seems he used the ship's call to Cobh as an opportunity to jump ship and see his family. Fr Frank Browne, a Jesuit priest, also came ashore, though, unlike Coffey, he did so reluctantly. His uncle Robert Browne, the Bishop of Cloyne, had paid for him to travel from Southampton to Cobh, but while on board an American couple had volunteered to pay the remainder of his fare to New York. Fr Browne sent a telegram to his Superior requesting permission to stay on the *Titanic*. The reply was succinct: 'Get off that ship.' He did so, an act which saved both his life and the hundreds of photographs he'd taken on board recording the ship's journey from Southampton to Cherbourg to Cobh. Some of Browne's photographs can be seen at the Titanic Experience by peering through little peepholes along the corridor, which give visitors the impression that they are getting a surreptitious glimpse of life aboard. James Cameron used several of Browne's photographs as inspiration when filming *Titanic*; one scene in the film showing a young boy playing with a spinning top on the deck of the ship almost exactly replicates a photograph taken by Browne.

At the Titanic Experience every visitor gets a boarding pass in the name of one of those 123 passengers – it was the first of many boarding passes, and alternative identities, I assumed while researching this book. I was Katie Mullen, a twenty-two-year-old from Cloonee in Longford, travelling third class, and on my way to my sister in New York. Misha was Thomas Smyth, a twenty-six-year-old Galwegian and third-class passenger. Misha glanced at my ticket and said, 'You might be lucky, 'cause you're a woman. I'll definitely drown.' 'How do you know?' I asked. 'Well, I'm a man and I'm in steerage.' Misha was a veteran of the Titanic Experience, having visited previously on a school trip. It turned out he was right – Katie Mullen did survive, and Thomas Smyth did not. Statistically, Smyth had very little chance, for 87 per cent of all third-class male passengers drowned. By contrast, 75 per cent of all female passengers on the ship survived. One of the reasons so many third-class passengers died is that there was no way for them to easily access the lifeboats – a complicated system of corridors and stairs designed to stop passengers from different classes mingling made it difficult for third-class passengers to reach the decks. Even if they had made it to the boats, there were a mere twenty lifeboats on the ship – enough for 1,178 people if all had been filled to capacity. There were over 2,200 people on board. The White Star Line was breaking no laws – indeed, they had more lifeboats than they were obliged to carry. In the event, lifeboats were launched far from full. Following the *Titanic* disaster, ships were obliged to have sufficient lifeboats to accommodate all passengers and crew.

Our tour group was brought out on to the balcony where the first-class passengers would have waited before being taken to the now dilapidated 'heartbreak pier' to board the tenders that took them out to the ship. I got no sense of their journey through the building from ticket desk to departure. I think that was partly because below us there were lots of people sitting in the sun drinking pints (the building is half exhibition, half pub), and partly because nobody in Cobh in April 1912 could actually see the *Titanic*, for she was moored out beyond Spike Island. After we stepped in from the balcony, we walked through reconstructions of some of the cabins. The first-class 'stateroom' cabins had windows, oak-panelled walls, a large four-poster bed and several occasional tables and chairs, while the second-class cabins had mahogany bunk beds and a washbasin and mirror. The second-class cabins were so comfortable they were the equal of many other ships' first-class offering, and even third-class passengers had mahogany bunk beds. On the other hand, Misha and I agreed that if we were among the 698 third-class passengers having to share two baths between us, we would probably have remained unwashed throughout the voyage.

Of the 123 people who boarded the *Titanic* in Cobh, fourteen were from the small parish of Addergoole, which sits in the shadow of Nephin in County Mayo. Eleven of them drowned. The parish, of fewer than 2,500 people, was desolate when the news reached it. Everyone knew, and many were related to, those who drowned. In the parish village of Laherdaun there is a *Titanic* Memorial Garden with a bronze sculpture shaped like the bow of the ocean liner and four bronze figures walking towards it. I visited at dusk one evening and discovered that while, by day, people visit to remember those who lost their lives at sea, at night it's a place where teenage couples, perhaps inspired by the film, come to enact their romantic fantasies.

The sinking of the *Titanic* isn't the only maritime disaster with a connection to Cobh. On 7 May 1915, the First World War came close to the town when the RMS *Lusitania*, a passenger ship en route from New York to Liverpool, was torpedoed and sunk by a German submarine off the Old Head of Kinsale. The ship sank in nine minutes, and almost 1,200 of the nearly 2,000 passengers and crew lost their lives. The survivors were taken to Cobh, and 150 of those who died were buried in the Old Church Cemetery just north of the town.

It was hot when I climbed up the steps to visit Cobh Museum, housed in a former Presbyterian church with a panoramic view of Cork Harbour and Haulbowline and Spike Island. The church is packed with traditional museum cases, the kind that you lean over and peer into, many containing objects associated with the *Lusitania*.

As part of the recovery operation after the *Lusitania* was torpedoed, officials took note of any personal objects or unique features on the bodies that were brought ashore. In an attempt to identify the bodies, Cunard (the shipping line which owned the *Lusitania*) later published these details in the *Record of Passengers and Crew*. Two entries are particularly poignant: 'Male baby, 12 to 18 months, round chubby fat face, hair inclined to be red, small short nose . . . wore white cotton bodice . . . blue cotton overall fastened . . . with white buttons . . . Buried Queenstown, May 17th' and 'Female, 23 or 24 years, round full face, nose broad at top, full eyes, fresh complexion, good plump build, good teeth . . . Wore woollen singlet with lace collarette, dark blue jacket . . . wedding ring; silver enamel brooch, English and French flags crossed. Buried Queenstown, May 18th'. Displayed alongside these lists are some of the personal items mentioned there. Some are mere trinkets, others valuable pieces of jewellery, but every one of them seems priceless now, for they are all that remains of those who died. There's a swallow brooch studded with blue stones and engraved 'coo-eee', a wedding ring, some glasses, some blue enamel cufflinks and a plain gold ring with 'S.H.' on it. These were precious, loved items, some taken off bodies before burial in the hope of identifying the dead. What makes them particularly heartbreaking is that these items were never claimed by any families and their owners were buried without ever being identified.

Many artefacts assume the qualities of a religious relic, particularly when they are associated with a dead hero, but it's rare that an object is seen as a relic from the moment it's collected. John Law, a sergeant with the Royal Engineers, was one of those who tried to rescue passengers who had been on the *Lusitania*. He pocketed a hard tack biscuit from one of the lifeboats that arrived in Cobh and wrote a note to his mother on it: 'A relic of S.S. Lusitania sunk by a German Submarine off "Old Head of Kinsale", Friday May 7th . . . taken out of one of her boats Sunday 9th May 1915 by J. Law, Templebreedy'. On the reverse he instructed her in block capitals to 'KEEP THIS FOR A SOUVENIR'. Both a relic and a souvenir. A fitting object for display in a museum that is also a church.

There is another *Lusitania* exhibition in the signal tower at the Old Head of Kinsale. After a week of sunshine, I arrived in wet and blustery conditions and was, unsurprisingly, the only visitor there. The signal tower was built in 1804 and is one of eighty-one constructed as part of a series of defences designed to prevent an expected Napoleonic invasion in the early nineteenth century. At least from a distance, the tower looks similar to the tower houses that dotted the country in the fifteenth and sixteenth centuries. The signal towers were not as heavily fortified as the Martello towers which were built around the same time, and so could not withstand cannonfire, or any sort of sustained attack.

The signal tower has been restored and is covered in twenty thousand slates (as it was originally), intended to help the tower withstand the brutal storms coming in off the Atlantic. They were being sorely tested on my visit as the wind whipped around the building in huge, enveloping gusts. Panels describe the history of the tower and the many, many wrecks in the area, including the *Hercules von Barth*, which was wrecked with the loss of eleven lives in 1874. Two bollards (iron posts used to secure ropes) from the wreck are on display. But the signal tower is primarily a memorial to those who were killed when the *Lusitania* was torpedoed. The display largely consists of copies of newspaper articles, photographs of various ceremonies and a few small pieces from the ship which have washed ashore, including part of a step, a fragment of a window and – more touchingly – the sole of a shoe. Outside, there is a lifeboat davit (the crane used to lower lifeboats from ship to sea) from the *Lusitania* which was caught in a fishing net in 1965. Unlike the *Titanic*, the *Lusitania* had enough lifeboats for all passengers and crew, but the ship sank so rapidly that only six of the forty-eight were launched.

Not content with one memorial, the site has several – the davit and the memorial garden which houses 'The Wave' (a twenty-metre-long bronze sculpture engraved with the names of all the passengers and crew) as its centrepiece. Slightly away from the tower stands a third memorial – a metal image mounted on two poles. At first glance, it seems to be a photograph of the *Lusitania* at sea, but when I moved closer to it I realized that the image had been perforated many, many times – in fact, 34,513 times. When you look at the image and through it out to sea, it seems for a moment that the *Lusitania* is once again riding the ocean waves and about to cruise peacefully past the Old Head of

Kinsale. It feels as if a moment has been captured, just before disaster strikes.

The final wreck of the Cobh triptych is the *Aud*. In October 1914, Sir Roger Casement travelled to Germany hoping to raise an Irish brigade which would participate in a rising against British control in Ireland. He failed to enlist troops but in February 1916 the German admiralty agreed to send a shipment of arms to help the rebellion that was being planned for Easter. On 9 April the *Aud* set sail from Lübeck with twenty thousand rifles, ten machine guns, explosives and ammunition, all destined for the Irish Volunteers. The German naval crew sailed under the neutral flag of Norway and planned to land the weapons at Banna Strand near Fenit, a small village on the north side of Tralee Bay, County Kerry. They expected to meet Casement there, but confusion about how the rendezvous was to take place meant that the *Aud* was left sailing around Tralee Bay waiting for a signal that never came, and on Easter Saturday, 22 April, the ship was intercepted by the Royal Navy. She was escorted to Cobh, but at the mouth of the harbour the captain, Karl Spindler, ordered his crew to put on their German naval uniforms, hoist the flag of the German Navy and detonate the explosives in the hold. They did so, sending the *Aud* and its cargo of arms to the bottom of the sea. Spindler and his crew escaped on lifeboats but were picked up by HMS *Bluebell* and taken to Spike Island, where they were briefly imprisoned. They expected to be executed, but instead they were transferred to England, where they were imprisoned for the remainder of the war.

The story of Casement and the *Aud* is hard to tell in museums, for there are few artefacts. Diving for salvage from the wreck is hazardous (and generally illegal) because of the amount of un-exploded ammunition on the seabed and, where the story is told, museums usually have nothing more than a handful of bullets or rusted rifles on display. However, there is a small but excellent exhibition about Casement at Kerry County Museum in Tralee, which includes a 'treasure map' drawn by Casement after his arrest, showing the location of gold and silver coins, a pistol and a pair of binoculars he buried after landing at Banna Strand. If only we'd had a treasure map when we were searching for treasure in Smugglers' Cave.

The scuttling of the *Aud* is the best-known maritime tale associated with the 1916 Rising, but there is one more that I often think about. In July 1917, Muriel MacDonagh went for a swim in Skerries. MacDonagh was the widow of Thomas MacDonagh, one of the signatories of the 1916 Proclamation, who was exe-cuted in Kilmainham Gaol in the aftermath of the Easter Rising. It was thought that she was swimming towards Shenick Island and rumours abounded that she was planning to raise the Irish flag on the roof of the Martello tower. In fact, this was unlikely, for the island is accessible by foot at low tide. The following day her body was washed up near Loughshinny. Her death would be a footnote in Irish history if it weren't for a box of Skerries sea-shells. Before her swim Muriel had collected seashells from the shore with her young daughter Barbara. Barbara kept the shells, and when the MacDonagh family donated their archive to Kilmainham Gaol they stipulated (somewhat bizarrely) that the box of shells must be put on display. And they are. I have visited the prison museum many times and on every trip I look at them and think of Muriel MacDonagh, wandering along the south strand, carefully choosing shells for her four-year-old son, and the maritime tragedy that was about to befall the family, leaving two young children orphaned.

Sometimes I've been surprised how blasé tour guides can be. But the passage of time does dilute the horror of the lived experience. We can't expect guides to spend their days ruminating about how horrific it must have been to die slowly and painfully at the bottom of an oubliette while a raucous banquet raged above, or the terror of facing the hangman's noose, or the panic felt as a ship slowly sinks. They have to try to find levity and humour in the past, particularly in the distant past. At times on my travels I've worried that I've become immune to the sorrow that death and tragedy has caused. In part, my camera and my notebook have become shields, used to deflect emotion by forcing me to try to make sense of what happened, encouraging me to seek a clear narrative rather than respond directly to the stories of suffering before me.

But there were moments on the journey where being drenched in sadness and loss can 'catch the heart off guard and blow it open', as Heaney put it in 'Postscript' (although he wasn't talking about grief). One of these occurred not at a museum, nor when I was thinking about the distant past, but while I was standing on a beach on the Mullet Peninsula in County Mayo admiring the Inishkea islands, glinting in a rare flash of sunlight. To the left and beyond I could see Blackrock Island, rising like a pyramid from the sea. It is a beautiful sight, and one now tinged with sadness.

In March 2016 the Irish Coast Guard helicopter *Rescue 116* crashed while on a rescue mission. The four crew on board were killed. There is something unbearably poignant about the death of people who died trying to save others. I stood looking out at the beautiful yet unforgiving Atlantic with my friend Agatha. Her brother-in-law, Vincent Sweeney, the lighthouse keeper at Blacksod, had sounded the alarm that March night and Agatha herself was part of the volunteer RNLI crew who had spent hours, days, weeks trying first to rescue, later to recover, the lost crew. This is a tight-knit community. I had followed this story through the newspapers, radio and television news. I read about the hundreds of volunteers, the sandwiches made, the tea poured, the beds offered up. But it is only by being there that you can sense the magnitude of the loss, not only on the family and friends of those who have died, but on a small community surrounded by the tempestuous ocean on three sides.

Famine

&

'The sleek dogs of Arranmore were my horror'

ASENATH NICHOLSON,

Annals of the Famine in Ireland

N

6 Doagh Famine Village

Dunfanaghy
Workhouse

National Famine
Memorial

1 Strokestown Park

Monasterboice

Doolough Famine
Memorial

3
**Irish
Workhouse Centre**

Kilmainham Gaol

Cabbage Gardens

Donaghmore Workhouse ■

4
■ MacDonagh Junction
Shopping Centre

Slea Head 5
Famine Cottages

Cork City Gaol ■   ■ Kindred Spirits sculpture

2 Skibbereen Heritage Centre
and Abbeystrewry Cemetery

Spike Island

I lifted the spoon to my mouth and hesitated. Three hours ago, this had seemed like a good idea, but now I was starting to regret it. As I'd travelled around the country visiting exhibitions about the Famine, one of the very few artefacts that appeared time and again was the soup pot – a large, black cast-iron cauldron. They're scattered across the country, not just in museums but everywhere, thrown about in yards, repurposed as flowerpots. My friend Danielle has two in her garden in West Cork. It's no surprise that so many pots still exist, for they were durable and plentiful. At the height of the Famine, in July 1847, nearly three million people were dependent on a bowl of soup every day. Thousands of pots were required, each one capable of holding hundreds of litres of soup. I wanted to know what Victorian philanthropy tasted like and so decided to make some famine soup. I settled on using Alexis Soyer's recipe, at least in part because he was the only chef to proudly lay claim to a 'famine soup'. Surely, I thought, it must have been the tastiest and most nutritious.

Alexis Soyer was one of the first celebrity chefs. A

flamboyant Frenchman, he was chef at the Reform Club in London and regularly cooked for the rich and famous. On the morning of Queen Victoria's coronation in 1838 he rustled up breakfast for two thousand guests. Like modern celebrity chefs, he sold his own range of kitchen products, sauces and recipe books. His breaded lamb cutlet is still served at the Reform Club. Long before Jamie Oliver attempted to revolutionize school meals in Britain, Soyer claimed he could provide cheap, nutritious soup on a large scale for the starving Irish, and in April 1847 he set up his 'model' soup kitchen at Croppies Acre in Dublin. It was a large wood-framed tent with a fabric roof. Long benches and tables were installed around the edge while Soyer's cooks stood in the centre, stirring their pots like a coven of philanthropic witches. The grand opening of the soup kitchen was an incongruously glittery affair. Soyer was there, an elegant figure, with his extravagantly embroidered waistcoat and his trademark red velvet hat cocked at a jaunty angle, accompanied by the Mayor, the Lord Lieutenant and the Dublin elite. Society ladies and gentlemen who missed the launch could visit on other days, for spectators were encouraged to come along. The well-heeled paid five shillings to amble about, peering at the poor as they ate their soup using spoons attached by chains to the table. This was a dining experience far removed from the decadent glamour of the Reform Club – at Croppies Acre, one hundred people were fed every six minutes. At maximum capacity, the soup kitchen was feeding over eight thousand people a day. I imagined the frantic clatter as they tried to swallow as much soup as they could, their desperate hunger, their scalded mouths.

The soup kitchen may have been well intentioned, but the decision to admit an audience was also exploitative. The *Freeman's Journal* compared it to paying to watch feeding time at the zoo: 'Five shillings each to see paupers feed! – five shillings each to watch the burning blush of shame chasing pallidness from poverty's wan cheek! – five shillings each! When the animals at the Zoological Gardens may be inspected at feeding time for sixpence!' Soyer appeared largely immune to the criticism (partly because he had flitted into Ireland for only a few days and while the Irish press were critical, his efforts were largely lauded in the British papers) and he continued to provide charity, though he restricted himself to raising money rather than setting up more soup kitchens, for all the government soup kitchens in the country were shut down by the autumn of 1847 (and almost all the privately run kitchens had closed by the summer of 1849). And he continued to spin famine into fame. *Soyer's Charitable Cookery, or the Poor Man's Regenerator* was published in 1847, and a penny from every sale was given to charities working with the poor. The book contained twenty-three 'nutritious' recipes, as well as a guide to creating a 'model' kitchen. Although few of those most in need would have had the money to buy the book, the literacy to read it, or the capability of obtaining the ingredients for dishes, which included Oyster Porridge, Curry Fish, Meagre Pea Soup or Cabbage Stirabout, the money raised may have been of some benefit when dispersed through Soyer's 'charitable committee'.

There's no name for Soyer's soup – it's simply 'Receipt [sic] No. 1'. The recipe called for two gallons (nine litres) of water, but since I don't own a famine pot and wasn't planning to feed anyone but myself and a less than enthusiastic Al, I divided Soyer's quantities by six. My soup consisted of:

- 10g of beef
- ⅓ of a turnip
- ⅓ of an onion
- 2½ sticks of celery
- ⅓ of a leek
- 40g pearl barley
- 40g plain flour
- 2 teaspoons of salt
- a tiny pinch of brown sugar
- a tiny amount of dripping
- 1.5 litres of water

Preparation was straightforward. I chopped the vegetables and put them in the pot with the tiny scrap of meat (10 grams of beef sits easily on the pad of my thumb). I let them soften before adding the flour, barley and water and then left it to simmer for more than two hours. As I waited for the soup to cook, I half (but only half) regretted not paying €25 for the miniature 'replica famine pot' I'd seen for sale at an exhibition, since it was the ideal size for a bowl of famine soup. By the time the soup was ready it had the colour and consistency of wallpaper paste. I ladled out two portions and we dipped our spoons. Al pulled a face. It was edible – grey sludge with a faint hint of turnip, reminiscent of the worst sort of school dinner. Still, the taste was largely beside the point. What mattered was the nutritional value. Sadly, in this respect, too, the soup fell short – each serving provided less than seventy-five calories. Today, an average woman needs about two thousand calories per day. Nutritionally, and in every other respect, Soyer's soup was useless, and certainly not worth the energy expended walking miles to get. Coincidentally, as I ate my soup I read a newspaper article which claimed that a spoonful of mashed potato was just as good for providing an energy boost as the energy gels marathon runners use. One potato, had it been available, would have provided more nutrition than a bowl of Soyer's soup. But that, of course, was precisely what the starving people lacked.

Despite what many people were taught in school, Sir Walter Raleigh did not introduce the potato to Ireland in the late sixteenth century – though for anyone interested in vegetable cultivation in Ireland, it is said that Cromwellian soldiers introduced cabbages in 1649, and there is a small park close to St Patrick's Cathedral in Dublin called 'The Cabbage Gardens', where they were first grown. Raleigh may not have brought the potato, but it is certain that the plant was being grown in Ireland by the early seventeenth century. Although far from its native South America, the potato thrived in the mild, damp conditions prevalent in Ireland and by the late seventeenth century it had become a staple of the Irish diet, particularly for the poor. There were a number of reasons for this. The potato is easy to grow; it thrives in poor soil that can sustain few other crops. It's nutritious and requires no additional processing or extra ingredients to make it edible – just plant it, leave it, dig it, cook it. The down-side of the potato is that it is susceptible to disease, particularly *phytophthora infestans*, more commonly known as blight – an air-borne fungus that from time to time could, and did, wreak havoc.

I say 'from time to time' because there is more than one Irish famine, though the only one most people have heard of is the 'Great Irish Famine' (or the 'Irish Potato Famine', as it's known in the United States). Perhaps to acknowledge more than one diminishes the impact. One story will suffice. But the fact remains that the population of Ireland was devastated by fam-ine twice in a century. The famine of 1740–41 was calamitous: historian David Dickson has estimated that up to a fifth of the population died in just over a year. An account written from West Cork noted that '. . . the dreadfullest civil war, or most raging plague never destroyed so many as this season'. But it's the Famine of the 1840s that has assumed a central place in the Irish popular imagination and identity. There are several rea-sons for this: it's more recent, it lasted longer, there are far more records available, and many of those who suffered and survived passed down stories of An Gorta Mór (the Big Famine), not just in Ireland but abroad. The Irish diaspora, created in part by this famine, built communities based on memories of hunger, evic-tion and exile.

According to the census, Ireland had a population of 8.2 mil-lion in 1841, though the figure by 1845 was probably close to 8.5 million. The census of 1851 recorded a population of 6.5 million. An estimated one million people had died from starvation, mal-nutrition and disease, and a further million had emigrated.

Much of the hardship was caused by the failure of the potato crop. The potato formed the key element of the diet of the very poorest people – primarily the landless labourers and impoverished small tenant farmers who lived in tiny mud cabins that dotted the countryside. The spread of 'the blight' through the country destroyed the potatoes in the soil but left the leaves and stalks largely unaffected, a particularly cruel detail that meant you couldn't tell whether your crop was ruined until you dug it up and the sickly-sweet smell of decay wafted across the field. The blight rampaged throughout much of northern Europe, but it was particularly devastating in Ireland, where so many were dependent on so little.

As with many other events in Irish history, the Famine is still contentious. While everyone is agreed that the impact was catastrophic, who (or what) should be blamed for it has been hotly debated. The science is beyond dispute, the politics is not. Two governments straddled the period of the Famine – the first led by Robert Peel (1841–6), the second by John Russell (1846–52) – and while both made attempts to alleviate the impact of the potato crop failure, they were inadequate to deal with the scale of the crisis. The amount of government aid, the appropriateness of the relief schemes and the manner in which they were applied has been the subject of much discussion, academic and otherwise. The key government-sponsored programmes were public works, soup kitchens and workhouses. At the height of the public works programme in the spring of 1847, nearly seven hundred thousand men (and some women and children) were employed to build roads, walls, piers and bridges to earn money to buy food. The public (and some private) projects are often the only visible reminders of the Famine – mute, unmarked memorials to those who suffered: a stone wall across the Burren's barren hillside, the star-shaped moat at Birr Castle, the pier at Portnablagh in Donegal. On the public works scheme, the pay was poor and, even with money, food was hard to come by. The workers weakened, the pay became piece work and food prices rose. People died as they worked and their families starved. The soup kitchens, begun by the Quakers and then embraced by the government, fed three million a day in July 1847 but were shut down by that winter, through a combination of the misguided belief that the harvest would be better that year, the difficulties associated with administering them and the decision to offload responsibility for outdoor relief on to the local Poor Law unions and the workhouses. By early 1848, there were nearly eight hundred thousand people receiving outdoor relief – rations of food – funded by local unions, which meant that the areas already most severely impacted by the Famine had to spread meagre resources even more thinly. The unions also funded the local workhouses, which were always the place of last resort for the ill and starving. These hulking buildings were cold, cruel institutions, often heaping shame on, rather than offering succour to, those who reluctantly begged for entry.

A decade after the Famine the Young Irelander John Mitchel wrote, 'the Almighty, indeed, sent the potato blight, but the English created the famine'. His argument that the British government, supported by landlords, was behind a deliberate and sustained plan to exterminate poor Irish Catholics found a ready and eager audience. Since then, versions of this assertion have been bandied about in books, newspaper articles, speeches and on the gable ends of houses in Belfast, where murals about the Famine appeared with the powerful (if inaccurate) message that the Famine was 'Ireland's Holocaust'. As recently as 2012, the journalist and author Tim Pat Coogan perpetuated the genocide argument in his book *The Famine Plot*, largely through a wilful misreading of the evidence. No historian would agree with him, but that doesn't stop the idea from floating around in the ether like a set of nasty spores – google 'Irish famine' and 'genocide' and you get over six hundred thousand hits (a few results link to articles arguing that the Famine wasn't a genocide, but many more assert with great confidence and little evidence that it was). A key part of the United Nations definition of genocide is that the acts that take place must be committed 'with intent to destroy, in whole or in part, a national, ethnic, racial or religious group'. The key word here is 'intent'. Governments, organizations and individuals could and should have done a lot more. We can point the finger at empire, berate the British governments for not taking sufficient action and condemn the greed of some landlords. But there is no evidence of any grand plan to destroy millions of Irish men, women and children. Still, the fact that it wasn't a genocide shouldn't detract from the horror of it. Many people could have been saved, many people who could have taken positive action did not, or they took action that caused homelessness, starvation and death.

In Ireland, there was little academic interest in the Famine for at least a century, in part because university history departments focussed on much earlier periods – examining something that had happened less than a century ago was regarded as virtually current affairs. To mark the centenary of the Famine, the government commissioned *The Great Famine: Studies in Irish History* (though it was not published until 1956). For the most part, it's worthy, but dull, an attempt to drain the emotion from a topic that is full of it. There followed a series of other worthy but dull books (with the honourable exception of Cecil Woodham Smith's *The Great Hunger* [1962], which remains engaging, if at times inaccurate). The suffering endured by those who lived and died during the Famine was largely excised from academic texts as historians tried to be scientific about the catastrophe, but it's also true that the voices of the victims were hard to find. But recent historians have grappled with these challenges and now there is a better balance, as a series of scholars including Cormac Ó Gráda, Peter Gray, Breandán Mac Suibhne and Margaret Kelleher have broadened our understanding of the Famine considerably.

What followed the Famine was silence, perhaps in part because, as Ó Gráda has suggested, the worst-affected areas were too traumatized to speak of what had happened, and perhaps nobody wanted to listen. As a result, for many years the challenge of balancing accurate research with a sense of the suffering endured by ordinary people seemed hard to achieve. If the only stories told were those that implicated decision-makers in London and landlords and their agents closer to home, that might be one thing, but digging deeper might illuminate all sorts of truths that might be safer left in the shadows. The mirror might not reflect what we want to see. We are all victims as long as you don't look too hard. But the harsh truth is that acts of cruelty during the Famine were not restricted to landlords: in March 1847, near Rosscarbery in County Cork, two young children had their throats slit by a neighbour who then stole the small bag of oatmeal and the loaf of bread that were in the house. And there were those who did not resort to violence but who thrived at the expense of family, neighbours and friends. In a recent memoir the author and screenwriter Maggie Wadey noted that her great-great-grandfather Daniel Dunne prospered during the Famine through 'caution, hard work and ruthless opportunism' when he expanded his farm in north Tipperary by acquiring the land of a neighbour he had been instrumental in having evicted and, later, more farms previously occupied by dead or emigrant neighbours.

The magnitude of the loss that occurred during the Famine is almost impossible to imagine. What is a thousand, ten thousand, a hundred thousand, a million? The numbers are impersonal and abstract. How do you tell a story so large and so calamitous while retaining a sense of the human scale that ensures an emotional response? For me, statistics have little impact; it's the small details that are the most moving. During the Famine, Asenath Nicholson, a Protestant American, travelled through Ireland dispensing aid and Bibles, hymns and religious tracts, to those who would take them. In mid-1847 she was on Arranmore Island, just off the coast of Donegal, and innocently asked her companions why the dogs looked so fat and shiny when the people had no food. Their pause, Nicholson recalled, was enough. 'If anything were wanting to make the horrors of a famine complete,' she wrote, 'this supplied the deficiency . . . the sleek dogs of Arranmore were my horror.' Nicholson's dawning realization that the 'sleek dogs' were thriving because they were eating the flesh of the dead is, for me, one of the most evocative and distressing images of the Famine.

At first glance, the museum at Strokestown appears an unlikely, even inappropriate, site for the National Famine Museum – a landlords' stately pile, the home of the very people who have long been condemned in popular memory for doing little or nothing to alleviate the suffering of their tenants. Strokestown has a population of just over 800 and the widest main street outside of Dublin. At the end of this rural boulevard visitors drive under a large triple-arched gothic gateway and down a sweeping avenue, where Strokestown Park House appears on the left, a three-storey Palladian mansion flanked by two-storey wings. The house was designed in the 1730s by Richard Castle, the architect responsible for many fine Georgian houses including Westport House, Russborough House and Belvedere House. In 1979 Jim Callery, a local garage owner, bought the Strokestown Estate from Olive Packenham Mahon, whose family had lived there since the mid-seventeenth century. When Callery inspected the house he discovered fifty-five thousand documents, many of which related to the Famine, and he was determined to use the documents and the site to tell the story of Famine.

The local story most closely associated with Strokestown concerns the murder of Major Denis Mahon, the owner of the Strokestown Estate. As conditions deteriorated for tenants across

Roscommon, there was growing violence towards landowners led by agrarian organizations and tenants and a hardening of attitudes among the landowners. Mahon was far from the worst of the landlords and he had, initially at least, been reluctant to evict tenants who could not pay their rent. However, following discussions with his land agent, he concluded that evicting tenants and sending them to the Roscommon workhouse would cost his estate £11,000 per annum (as the estate partly funded the workhouse), whereas for a one-off payment of less than £6,000 he could send three thousand people to Canada and remove the problem permanently. In early 1847, Mahon began an assisted emigration scheme which saw more than 275 families (1,490 people in total) provided with tickets to Canada. But these tenants did not choose to go, they were forced to emigrate by Mahon who selected those 'of the poorest and worst description who would be a charge on us for the poor house or for outdoor relief'. Many of those evicted were already ill and the dreadful conditions on board the ships charted by Mahon exacerbated these illnesses, with typhus spreading rapidly through the passengers. Four hundred and ninety-six Strokestown tenants boarded the *Virginius* at Liverpool and during the voyage 267 died and a further 180 fell seriously ill. Of the 1,490 who left Roscommon, over 700 died at sea or when they reached the quarantine station at Grosse Ile, an island in the Gulf of St Lawrence in Quebec.

In addition to those dispatched to Canada, Mahon also evicted hundreds of other tenants, and at Mass on 1 November 1847 the local priest, Michael McDermott, allegedly delivered an inflammatory sermon that Mahon's actions, claiming that 'he is worse than Cromwell and yet he lives'. The following day, as Mahon travelled home from Roscommon Town, where he had attended a Board of Guardians meeting (at which he had tried to secure additional relief for his remaining tenants) he was shot and killed. Ten months later two men, James Commins and Patrick Hasty, were hanged for the murder. They may well have been guilty, for although Hasty was not a tenant on the Mahon estate, he was a member of the Molly Maguires – a secret society that used violence to highlight land issues and tenant rights. Guilty or not the police were determined to charge the men with murder and pressured a number of people to give evidence against them.

The museum effectively tells the parallel stories of landlord and tenants on the Strokestown estate. With the exception of the room imaginatively recreated as a Victorian dining room the colour scheme is dark and oppressive, mirroring the despair felt by many in the 1840s. On display are many documents – incredibly valuable resources, but in a museum setting it's hard to generate emotion from yellowed paper with hard-to-read handwriting. The exhibition navigates this problem by playing a recording of the Cloonahee Petition alongside the highlighted text itself. The petition is a remarkable document from 1846 which vividly depicts the desperation of the tenants at Strokestown who were facing starvation and eviction.

There are few artefacts in the museum associated with the tenants. This is unsurprising for the poor leave little behind. There are soup pots, a foot iron used for cultivating potatoes and a small shard of mirror – about the size of a pocket notebook – set into a wooden frame. This reminds me of an inventory of the material wealth of the community in Gaoth Dobhair (Gweedore), Co. Donegal which was compiled by a schoolteacher in 1837, eight years before the famine. Between them, the population of about 4,000 people owned: 1 cart, 1 plough, 20 shovels, 32 rakes, 7 table forks, 93 chairs, 243 stools, 2 feather beds, 8 chaff beds, 3 turkeys, 27 geese, 3 watches . . . and one tiny mirror. One tiny mirror between four thousand people. They had no need of a mirror, one of life's small vanities, when it was a struggle every day to put food on the table and keep a roof over their heads. The list is a stark record of life on the margins, where one small change was devastating and where the loss of the potato crop could be fatal. Gazing at the shard of mirror, now a banal, everyday, object I find there is something evocative and moving about that remnant of lives lived in incredibly harsh conditions.

At the outset of my visit, as I crossed the courtyard from the visitor reception to the exhibition, I stepped on words carved into granite paving slabs – Dóchas and Uafás (hope and horror), Mishneach and Cruálacht (courage and cruelty). There was plenty of horror and cruelty in the Strokestown story and undoubtedly there was courage, though sadly I saw little that gave me hope.

As this project developed, I began to see misery everywhere. On my way to collect Misha from his home in St Luke's on the north side of Cork city, I noticed the green sign above the small post office. In English the sign just reads 'St Luke's', but in Irish the post office is at Crossaire na mBocht, the 'Crossroads of the Poor'. A few hundred yards from the crossroads is the romantically named 'Lovers' Walk'; in Irish, it's rather less appealing, for Siúl na Lobhar means 'Lepers' Walk'. Every street has a story, but that day the stories I was in search of were further west.

The beauty of West Cork deserves whatever superlatives are thrown at it. It is one of the most scenic parts of Ireland, replete with sandy beaches and rugged peninsulas. But this beauty masks a very dark past. At the height of the Famine, this part of Ireland saw almost unimaginable suffering. We passed opportunities to take turns that would have brought us to the magnificent beaches of Inchydoney or Owenahincha, instead focussing on getting to Skibbereen Heritage Centre and Abbeystrewry Cemetery to learn about the horrors of the Famine in the area. It felt strange on a very hot July day to be driving around West Cork chasing tales of torment while everyone else was in pursuit of the beach and ice-cream.

The Skibbereen Heritage Centre is housed in a former gas-works; the exterior is a cheerful mix of exposed stonework and brightly blooming flower boxes and hanging baskets. We arrived at the same time as a small party of Americans, who were, as they repeatedly told the very patient man at the ticket desk, 'coming home'. We listened as one by one they told him their 'Irish surname', which, in most cases, had disappeared after generations of marriage but often resurfaced as the first name of a son, daughter or grandchild. One of them confidently asserted that she was the first member of her family to come 'home' since they'd left during 'the Great Potato Famine of 1893'. As a historian who can't remember dates, I could only empathize.

The heritage centre opened in 2000, and it feels a little jaded now, cluttered with text and exhibition cases. The story, too, adds to the sense of being hemmed in. The focus of the exhibition is on local stories, of which there are many – it has been estimated that the Mizen Peninsula just west of Skibbereen lost 70 per cent of its population through starvation, disease or emigration during these years, and at the time Skibbereen and the surrounding area became synonymous with the Famine, not least because of the tireless work carried out by Dan Donovan, a local doctor and the first medical officer at Skibbereen Workhouse. Donovan is undoubtedly a hero, for he worked tirelessly, caring for the poor and the ill. The diary he wrote for the *Southern Reporter* (which was reprinted in major publications like the *Illustrated London News* and *The Nation*) outlined the suffering he saw every day and helped focus national and international attention on what was happening in the area. In one memorable article, Donovan described Christmas 1846 as 'the triumph of pestilence and the feast of death'. But no stories are as straightforward as we might like them to be, and while Donovan was railing against injustice and appealing for assistance, fifteen kilometres away his mother was evicting Jeremiah O'Donovan Rossa's family from the land they farmed in Reenascreena.

It was partly because of the attention that Donovan generated that the artist James Mahony arrived in Skibbereen in early 1847 determined to record with 'unexaggerated fidelity' the devastation caused by the failure of the potato and the inadequate response by the state. Mahony's illustrations for the *Illustrated London News* became, and remain, the defining visual images of the Famine. One of his most well-known shows two painfully thin children dressed in rags. The young girl is scrabbling at the bare soil in a vain search for food; the boy's hollow-eyed gaze stares beseech-ingly out at the viewer. The dissemination of Donovan's words and Mahony's images did have an impact, and aid was sent to Skibber-een and elsewhere, though it was not enough to do much more than alleviate the suffering of a few. Alongside thousands of indi-viduals and organizations, many journalists and newspapers also donated money for famine relief – the proprietors of the *Illustrated London News* donated £10, while both the *Observer* and *Punch* donated £50. The *Punch* donation is a surprise, as the paper fre-quently printed anti-Irish cartoons. Another paper with an anti-Irish bent, *The Times*, was actively hostile to any charitable efforts and urged readers not to donate. Despite the modest relief efforts in Skibbereen, there were certainly those who felt that the town received a disproportionate amount of attention and aid. In the late 1870s, James Daly, editor of the *Connaught Telegraph*, sar-castically praised Dublin journalists who had gone to Mayo to cover a Land League story. These journalists, he wrote, 'were never done poking at the famine pits of Skibbereen because there was a smart local doctor who wrote them up . . . Two hundred thousand people died of hunger in Mayo . . . and the world never said so much as "God be merciful to them."'

Mahony's illustrations and Donovan's words form a key part of the exhibition in Skibbereen, but one of the exhibits that caught our attention was neither artefact nor text. It was a hessian sack sitting on the floor with '55lb, 25kg' printed on it. A note beside it explained that twenty-five kilos of potatoes was eaten every day by a family of six. Misha bent down and picked it up, stumbling as he did so. Theoretically, providing twenty-five kilos of potatoes a day for a family might not seem much, but picking up a sack of them makes the task much more real. There's something immediate and shocking about the heft of it. Twenty-five kilos weighs two and a half times Ryanair carry-on. Imagine a family having to eat that much every day just to survive. And imagine how terrible it would be when that food was no longer available.

I made slow progress through the displays, determined to read every panel, and was quickly lapped by Misha before he disappeared. It's hard to catch and keep a teenager's attention, harder yet when it's gloriously sunny outside. Perhaps he was overwhelmed by the misery on display, or suffering from back strain from the sack of potatoes, or, I suspected, he was simply bored and moping in an alcove, waiting impatiently for me to be done. But I was wrong. After ten minutes, he returned, his eyes shining with delight, and hustled me over to a corner where two women were sitting with small speakers clamped to their ears. A woman wrapped in a dark woollen cloak appeared on the screen in front of them. I had no idea what I was supposed to be looking at until Misha pointed at two green buttons below the screen offering viewers a choice between two kinds of stories. The buttons were labelled 'Suffering' and 'Death'. 'That's your book,' said Misha, 'suffering or death.'

When the Famine becomes 'visible', it tends to be in places where it intersects with the institutions that did so little to relieve it: prisons and workhouses. To the British government (and many landlords), the Famine was a threat to public order and they were determined to quash any prospect of revolt. Unsurprisingly, this resulted in a surge in the prison population in the late 1840s, in part because of crimes of desperation which saw adults and children imprisoned for very minor offences, including stealing a 'copper coal scuttle', 'one pear from a garden' and, in one instance, 'handfuls of grass'. Some of the first convicts sent to the prison on Spike Island near Cobh to await transportation were those who in desperation broke the law. Edward Walsh, a teacher in the prison, wrote to his wife in 1847 that 'most of the convicts are persons . . . who were driven by hunger to acts of plunder and violence . . . I wept today in one of the wards when some of the people of Schull and Skibbereen told me the harrowing tales of their sufferings from famine, and the deaths – the fearful deaths – of their wives and little ones.'

The prison population was further swelled by the Vagrancy Act of 1847, which made begging and homelessness punishable by a month of hard labour in prison. Kilmainham Gaol had just over a hundred cells, most intended to house one prisoner, but in 1850 over nine thousand committals were recorded (the average number of daily inmates was 259). The prison inspector's report noted that 'numbers of these wretched creatures are obliged to lie on straw in the passages and dayrooms of the prison without a possibility of washing or exchanging their own filthy rags for proper apparel'. Kilmainham was far from the only prison that was overcrowded during the Famine: in jails such as Wicklow and Nenagh it was common for cells designed for one prisoner to house at least five. Some people were so desperate they actively solicited arrest in the hope of a meal and a place to stay. The quality of prison food did decline during the Famine, but all prisoners were fed. And it wasn't just adults: hundreds of children were also held in these jails. In 1847, ten-year-old James McCormack was sentenced to forty days' hard labour for stealing parsnips, and nine-year-old Winnifred Carroll was imprisoned for a week of hard labour for begging.

Those who weren't arrested often threw themselves on the mercy of the state and found themselves at the door of the local workhouse. Workhouses were a response to the growing poverty evident in Ireland a decade before the Famine. The 1836 Poor Inquiry Commission reported that almost two and a half million people were in such a state of poverty in Ireland as to require organized welfare schemes. Two years later, the Poor Law was passed, and by the early 1850s there were 163 workhouses spread throughout the country, usually close to market towns. In 1847, there were about 417,000 inmates; by the end of 1849, there were over 932,000. Though most of the workhouses continued to operate until the early 1920s, in popular memory they are closely associated with the Famine. Many of the harrowing stories told about workhouses are not necessarily famine stories, but they could be.

Workhouses were forbidding, gloomy places deliberately designed to act as a deterrent rather than a refuge, for the government believed that people would become idle on welfare. They were all designed by George Wilkinson and were very similar – two- and three-storey limestone buildings built along a horizontal H-plan. Segregation of inmates was the primary concern. When a family passed through the reception area, they were separated, women from men, children from their parents, boys from girls. There were separate dormitories, yards and workspaces. At the back of the complex there was an infirmary for the ill and a 'dead house' for the deceased. 'Wee Hannah' Herrity, who was an inmate in Dunfanaghy Workhouse in north Donegal, recalled the cruelty of the matron, who, when inmates died, would have their bodies dragged from the dormitory and 'you'd hear the head of the corpse cracking down the steps till it was put in the dead house below'. There was no comfort for the living either – the ground floor of most workhouses was earthen, not flagged, the plain stone walls in the dormitories and workrooms were whitewashed but usually not plastered, and there were no beds in the dormitories, only raised wooden platforms covered in straw bedding. The regime inside the workhouse was prisonlike. Residents were referred to as 'inmates' and obliged to wear uniforms. Every day was strictly regimented and inmates were obliged to work to pay for their keep. Since workhouse employment was not allowed to compete with businesses outside the walls, the work was designed to be 'of such a nature as to be found irksome'. This meant that, alongside cultivating the food eaten in the workhouse, many inmates picked oakum and broke stones – jobs more usually associated with prison.

Picking oakum was horrendous work carried out by both male and female inmates. The skin on their fingers and palms was shredded by small fibrous tendrils as they untwisted old rope so it could be tarred and used for waterproofing boats and ships (when that was no longer done, the oakum was used to fill mattresses in the prisons). Inmates' hands were soon covered with scores of small cuts which had no time to heal, given that they spent eight hours every day unpicking the rope. In Cork City Gaol, visitors can handle some unpicked rope. Misha and I rubbed the dry, tough fibres between our fingers and imagined how it would slice through the skin like hundreds of paper cuts. Stone-breaking was no better – hard, exhausting labour carried out by male inmates to create the scree used in building and road-laying.

More than half of the former workhouses in Ireland have been demolished and, of those that remain, about forty operate as medical facilities – St James's Hospital in Dublin, Naas General Hospital and St Finbarr's Hospital in Cork are just three of those that have incorporated former workhouses into their modern facilities; a handful are heritage sites open to the public, though the history of the workhouse isn't always the primary focus. I'd never really thought of farmers as sentimental hoarders, but they must be, for throughout my tour I came across farm machinery in the most unlikely of places, where their brightly painted colours seemed purposely designed to tempt small children on to their blunt and rusted blades. In Donaghmore Workhouse in County Laois, several of the old dormitories are littered with these agricultural antiques. It's a little jarring to see that the platforms where the poor once slept are now the resting place of a selection of ploughs, seed spreaders and butter churns.

Portumna Workhouse opened at the tail end of the Famine in 1852, and today it provides the most comprehensive account of life in the workhouses from the Famine onwards. Misha and I walked out to the workhouse from the town, tracing the path trodden by many of those who became inmates. Like so many of the workhouses, the building completely dominated the landscape on which it sat – an austere grey pile surrounded by high external walls. As we approached the entrance Misha noticed that the building looked lopsided, as if blinking. On one side of the main door there were large sash windows allowing light to flood into a room; on the other a row of much smaller windows was set about seven feet above the ground. The first room was for the well-heeled Board of Guardians so they could admire the garden and a distant view of the town. The second was a schoolroom, where the children would have had a mere glimpse of sky. We started our visit with a cup of tea with Steve Dolan, then the manager of the workhouse, though it seemed incongruous to sit eating doughnuts and chocolate biscuits while discussing malnutrition and disease. Once we had been fed and watered, Aileen, one of the tour guides, took us off to see the site.

Portumna Workhouse is largely intact and Aileen showed us dormitories, a schoolroom, the matron's room, the laundry building and the yards. As we moved from building to building we got a real sense of how oppressive the site was: the high boundary walls, the segregated yards, the freezing dormitories (some with one tiny fireplace at one end of the room – which we were assured was hardly ever lit). At the end of the regular tour Steve reappeared and asked if we wanted to see parts that are currently closed to the public. Of course we did. Ever since our first trip – to Cork City Gaol – Misha had been keen to get a look behind the scenes at the sites we visited. This was our first opportunity. We traversed an overgrown yard (the path I chose seemed to involve blundering through every available nettle and I rapidly regretted my decision to wear shorts). Judging by the number of nests, and birds flitting in and out of the decrepit buildings, they still had plenty of inmates. Steve strode up a set of stairs with no bannisters, and Misha followed. I dallied, distracted by the graffiti etched on the wall – someone had gone to the trouble of carving 'Elvis Presley' and his date of birth and death – but I caught up just in time to see Steve bounding across the one remaining floorboard like a gymnast on the beam. Either side of the floorboard was a long drop to the floor below. I hauled Misha back before he could contemplate following suit. I'd promised to return him in one piece.

A shopping centre might seem an unlikely location for a famine tour, but that's exactly what's on offer at MacDonagh Junction shopping centre in Kilkenny. The modern centre, all glass panels and shiny tiled floors, encircles the courtyard of the old workhouse. The woman on the information desk was a round-faced twentysomething who looked thoroughly at home amid the brand-name stores, beauty boutiques and hair salons. But a significant proportion of her job, given that most people in a shopping centre don't really need information, is to distribute the 'Kilkenny Famine Experience' AV handset, and she handed ours over with a long and eloquently delivered spiel about where to go and what we'd find.

Donning the headsets, Al and I headed for the Workhouse Square food court. It was warm there. What had once been a windswept courtyard is now an enclosed space with a thick glass roof, all sandblasted cobbles and free wifi, the tinkle of cutlery and the smell of toasted sandwiches, and the only trace of the howling wind that once tore across the yard came from the play area in the centre, which was blasting out 'Let It Go' from *Frozen*. As Elsa sang, 'The wind is howling', I stood underneath Plaque 1 and pressed 'play'. 'The workhouse,' said the narrator, as I gazed at the families eagerly tucking into lunch, 'was home to the lost and lonely of Irish society.' Two of the lost and lonely were John and Patrick Saul. Aged fifteen and thirteen, they were among the first to be admitted to the workhouse in 1842, after they were abandoned by their parents.

We followed the trail around the food court down the main aisle of the centre, stopping at intervals to look at a detail mentioned in the audio or watch a brief video on the handset. Near the entrance we sat on a bench, beside which a silver-haired man in his fifties had set up a stand for Acorn Insurance ('Win a holiday worth €4,000!'), while I learned about the scurvy routinely suffered by the inmates – a scurvy that could have been easily prevented by the contents of the fruit stands at Dunnes. We watched a short film which introduced Jonny Gerber, a bioarchaeologist who had worked on the site. Two thousand two hundred people died in the Kilkenny workhouse during the Famine, said Gerber, and 970 of them were buried in a mass grave because the local graveyards were full. It was this grave that was discovered in 2005 when building work began on the shopping centre. Most of the victims had died of typhus, tuberculosis or scurvy rather than starvation. The small branch of Boots to my left probably stocked enough medicine to save them all.

The tour ended outside at the small memorial 'garden' opposite LifeStyle Sports, a discreet little triangle of sculpture and grass where the remains of the dead have been reburied. There's an Ani Mollereau sculpture to John and Patrick as boys playing together, and it's here, listening to the last track on the AV guide, that we learned that John had emigrated to Australia. He turned up in Sydney in 1851 and seven years later was incarcerated in an asylum, suffering from mental health problems. What happened to Patrick is unknown. The audio ends movingly with 'Song on the Times', an English folk song from the mid-1840s:

> There's different parts in Ireland, it's true what I do say.
> There's hundreds that are starving for they can't get food today.
> And if they go unto the rich to ask them for relief,
> They'll slam their door all in their face as if they were a thief.

Though listening to stories about deprivation and starvation jars a little with the conspicuous consumption of the setting, the tour is an impressive achievement, well researched and sensitively presented, one of the most successful attempts to balance scholarship and emotion I have found on my travels. After many months of traversing the country, encountering a wide variety of attempts to convey the full horror of the Famine, it's here in Kilkenny, outside a branch of Eddie Rocket's in the rare August heat, awaiting news of the fates of two teenage boys, that the horror of the millions of individual personal tragedies is finally brought home. Surrounded by the hustle and bustle of slightly frenetic shoppers, I sit silently thinking of all the graves I've walked past and over, of the named and nameless dead. I look across and see a tear working its way down Al's cheek.

*

Workhouses are the most obvious visual reminders of the Famine. Even when they have been converted, they retain their distinctive institutional feel. But there are other reminders if you know where to look. The entire island is a camouflaged landscape of loss. Halfway up Slievemore on Achill Island, faded diagonal lines scored into the earth catch the light as the sun disappears into the Atlantic: lazy beds, a misnomer if ever there was one. These scars are reminders of cultivation, when the potato was sown high on the hills, higher than any other crop, often higher than the tree line. These beds were not an easy option but dug out of necessity and the desperate desire to feed a family.

Very few buildings and very few gravestones remain to commemorate the Famine dead. The markers of life and death are easily erased if you're poor. The mud cabins melt back into the earth, grass grows over the unmarked graves, there are few objects to keep and few people to treasure them. But sometimes absence tells the tale. As I drove back to Skerries one afternoon I spotted a brown sign for Monasterboice – an early Christian settlement best known for its round tower and two high crosses. I'd been dragged there as a child on numerous occasions, and in my memory it was in the middle of nowhere (in fact, it's about five minutes off the M1, a few miles north of Drogheda). The tenth-century West Cross, twenty feet tall and carved with biblical scenes, remains as impressive as I remembered it. But the lower portion of the cross is quite badly damaged, not – as I thought – from generations of cattle using it as a scratching post, but, according to the information panel at the entrance to the graveyard, from local people during the Famine chipping off a piece of the cross, a relic from home to bring with them when they emigrated. True or not, it shows how the folk memory of the Famine permeates everything.

Graveyards and cemeteries are some of the few places where the Famine is remembered. Art historian Emily Mark-FitzGerald has identified ninety monuments and memorials to the Famine erected between 1991 and 2013, and twenty-seven of these are at burial sites. During the Famine very few individual headstones were erected, and in some parts of the country so many were dying that thousands were placed in large communal burial pits, unmarked for many years. Indeed, not all were afforded the dignity of their own coffin. Some were piled on to carts in their rags without shroud or sheet, others placed in hinged coffins – something Dan Donovan referred to as the 'parish coffin' – a grimly efficient device. When the body was being lowered into the grave the base would swing open and the body would fall through, allowing the coffin to be lifted up and reused.

In good weather, the drive from Louisburgh to Leenaun is beautiful, the browns and greys of the bog rolling out until they meet the foothills of the Mweelrea Mountains and the Shaffrey Hills. The road undulates, and at the highest point of elevation you can see the dazzling Doo Lough in front. But on the day I went – a miserable, cold, wet February day – the 'black lake' was living up to its name, a dark, inky mass spreading across the valley. The hills were completely masked by low cloud and the wipers at full tilt struggled to keep the windows clear of lashing rain. In March 1849 several hundred people had staggered down this road to plead with Poor Law Guardians for more famine aid. Many of those who travelled were already ill; all were weakened by hunger. When they arrived at Delphi Lodge, twenty kilometres south of Louisburgh, to argue their case, they were made to wait until the Guardians had finished their lunch. All further aid was refused, and the starving hundreds were forced to make their way back to Louisburgh. On the return journey several of them died from exhaustion and exposure.

A small stone cross stands on the edge of the road close to the lough as a permanent reminder of that walk. A bitter wind swept freezing rain across my face as I read the inscriptions on the base of the cross and thought of the suffering of those who had trudged past this site barefoot and in rags. The cross, which was unveiled in 1994, is one of a number of famine memorials where the text links the suffering of the past with later suffering in other parts of the world. The cross commemorates not only those who passed by in 1849 but also the end of apartheid in South Africa. It was erected by Afri (Aid from the Republic of Ireland), whose patron is Archbishop Desmond Tutu, and since 1988 this charity has held a commemorative walk to remember those who suffered during the Famine and to raise money for those who suffer today.

And there are many famine memorials dotted around the country. John Behan's National Famine Memorial at Murrisk rests at the foot of Croagh Patrick, looking out at Clew Bay. His bronze ship has skeletons as its rigging, and the ridges on the side mirror the lazy beds of the hillsides. It's an evocative but also 'silent' sculpture, for unlike most memorials there is no explanatory text nearby and the tribute stands without comment, leaving the visitor to their own reflections. Such silent memorials are deeply ambiguous things. They say 'I remember', but they don't always say what. There's a more uplifting memorial in Midleton, County Cork, where nine twenty-foot-tall steel eagle's feathers are arranged in a circle to mimic the shape of a bowl. 'Kindred Spirits' was erected in tribute to the Choctaw Nation, who, despite living in atrocious conditions themselves, sent £170 to help alleviate the impact of the Famine in Ireland. This act of great generosity occurred barely a decade after the Choctaw had suffered enormous losses when they were forced on to the Trail of Tears by a US government determined to push them into what they had designated 'Indian Territory' (much of which is now Oklahoma). The Choctaw Nation's contribution wasn't the largest (that came from Queen Victoria, who gave £2,000, rather more than the £5 some of my friends were told about at school) but, symbolically, it was one of the most important, and their generosity is still remembered today. In May 2020, in the midst of the COVID-19 pandemic, the Navajo Nation was in desperate need of medical and food supplies and appealed for assistance. Tens of thousands of Irish people answered the appeal, not only because such suffering still has particular resonance in Ireland, but also in recognition of the compassion shown by another Native American tribe over 170 years earlier.

In 1841, half of those living in rural Ireland were living in one-room mud cabins. On land where the threat of eviction was high there was little incentive to build substantial dwellings and many cabins were constructed as lean-tos against a bog-face or a ditch. There was a saying in County Armagh that 'you can't cross a ditch or you'll fall down a chimney'. Almost all of those cottages have dissolved back into the landscape. But some stone cottages remain, and a handful can be visited. In Kerry, there are Famine-era cottages on the Slea Head drive out of Dingle, sandwiched between the beehive huts that form part of the unofficial *Star Wars* trail and a fairy fort signposted with cheerful cartoon sheep. At the Famine Cottages and Sheep Dog Trial Centre the sheep are real, a variety of tame creatures with coats thick enough for a Russian winter, along with several donkeys and a number of handsome ponies, all vying for the attention of the tourists strolling up the hill towards the cottages (because nothing prepares you for the horrors of rural starvation like feeding adorable livestock from a plastic tub). Having softened up visitors with this preamble, the self-guided tour began at the first cottage, which is not a cottage at all but a tiny stone 'hovel' once occupied by 'Pat', a peasant farmer evicted from his nearby house.

Al and I proceeded to the main house, a four-room cottage with a fireplace and a settle bed and the first famine pot I'd seen that came with a lid. A grandmotherly mannequin sat darning wool by the fire (it appeared that wildlife had run off with one of her feet) while a girl with unruly tousled hair and a missing tooth squatted in a hole in the wall. There's a room with a couple of iron bedsteads and a sponge mattress. The third room contained a model of the terrain, complete with labels whose misspellings somehow add to the anguish (one woman evicted from her home lived 'under a large border' until her death). We stumbled out through the side door, blinking in the sudden light, to a small beehive hut which at one point had been home to the family pig.

But why visit three famine cottages when you can see a whole village? Back when I began planning my tour of the country, numerous people told me that I had to visit Doagh Famine Village in Donegal, but none of them would tell me anything about it. They said it would 'ruin the surprise', though a friend's eleven-year-old daughter told me it was the best place to go to see Santa. The Famine and Santa. Of all the incongruities on my tour, this promised to beat them all. Unfortunately, I was going in September.

I had expected that the village might be a larger version of the houses we'd seen in Kerry. A ghost village, in short, in which most of the inhabitants starved or from which they fled, in which poor relief was minimal or to no avail, a village that was abandoned and left to decay and in which, even 170 years later, the sound of lamentation still seems to linger in the air. Ideally, it would be situated in an area notorious for the cruelty and indifference of its landowners, and each house, dating from the 1840s at the latest, would be meticulously preserved at the moment of its inhabitants' greatest distress. All this, it seemed to me, was what you'd expect of a famine village, but it is far from what we found in Doagh.

Doagh Famine Village is a self-contained quad: four rows of thatched whitewashed cottages, with brightly painted doors facing each other. Our tour began in the family's original house, in a tiny front room with a hearth and a bed, where Pat, a charismatic host, explained life in the house when he was a child and talked engagingly about folklore and superstition. As I began to wonder when we'd get to the Famine, he led us over to a second cottage, all set up for a demonstration of poitín distilling – complete with optional tasting. From there we staggered to the largest cottage of all, which had been set up for a wake. There was enough seating here for a coach tour, and at the top of the room on a small stage stood an open coffin with a body in it made out of what looked like plaster of Paris. On the chest of the deceased was a saucer of snuff for mourners to use. A cluster of manne-quins surrounded the body in various degrees of dishevelment, some tending to the corpse, others drinking beer, all set for an evening celebrating the dead. Above the corpse was a large plas-tic sign which read: 'Death is a debt we owe to nature, I must pay it and so must you.' Al, who was suffering flashbacks to my uncle's wake, loitered at the back of the room. The presence of death in this cottage promised to lead us neatly to the Famine, but no, here was more familial anecdote unencumbered by dates. As we reached the fourth cottage, Pat handed over to another family member and disappeared.

It was only after the tour had been going for an hour that we got the first whiff of potato blight. Led out of the quad into a hitherto undreamed-of part of the 'village', we were confronted by a long crescent of outbuildings, each one filled with a differently themed diorama. To our left, an eviction scene populated by mannequins played out in dumb show beside a 'fallen woman' living in a hovel. The mannequins all looked like they'd been recently released from a window in Dunnes and seemed understandably miffed to find that their new life involved standing outdoors looking mournful while clad in Granny's old clothes (period costume is not the village's forte). Our young guide left us to explore the rest of the complex on our own. It was at this point that we began to real-ize what had happened to Doagh Famine Village in the decades since one family had a simple idea. I'll put it concisely: they didn't leave it there. What is attempted at the Famine Village is nothing less than a history of Ireland, with snatches of family 'amusement park' thrown in for good measure. Some of the amusement, it must be said, is unintentional. In a parting aside, our young guide informed us that the full-scale Orange Lodge we were about to admire had elicited protests from the local republican community, in the wake of which the family had installed, in the interests of fairness and balance, a Republican Safe House on the opposite side of the path. The Republican Safe House turned out to be a network of rooms with false doors hidden in wardrobes and behind shelves, each one leading to another room decorated with lively murals of figures involved in the peace process. This attraction segued somewhat less than seamlessly into the 'Spooky Village', a series of ghoulish rooms in which macabre skeletons painted in fluores-cent yellow rattled loudly every two minutes.

Staggering valiantly onwards, we encountered the Traveller exhibition, the Presbyterian Meeting House and the Mass Rock. Inspirational quotes (often of questionable relevance) were scat-tered throughout the village – one, paraphrased from American motivational speaker Zig Ziglar, reads, 'You don't have to be great to start something, but you have to start something to be great.' I began to wonder if that quote had prompted the development of the village. On our way towards the exit I noticed a somewhat disjointed mannequin pushing a wheelbarrow containing a body wrapped in a sheet. Propped against the wheels was a sign which proclaimed, 'You may never have this day again ... so make it count.' For the mannequins, it seemed like a day not worth having at all, let alone again. Amid the sensory overload, the original concept of a famine village – austere, focussed and small – has been somewhat lost, but in a world of increasing professionalization it is – very occasionally – refreshing to be reminded what an entrepreneurially minded family can do armed with nothing more than a series of thatched cottages, a truckload of mannequins, some fluorescent yellow paint and access to the internet. While other sites do a better job of conveying the horrific reality of the Famine, if you are at a loss for entertainment in Doagh and have a predilection for a drop of poitín, Doagh Famine Village is for you.

DETOUR

A Day at the Park

It comes over you slowly, almost unnoticed at first. Then suddenly it's upon you: heritage fatigue, a little-known and rarely diagnosed complaint that comes with many symptoms:

- Tour-guide tinnitus
- Signage stress
- Panel blindness
- Diorama dizziness
- Battlement vertigo

A combination of more than three of these requires a very long sit-down or an escape to the countryside. We decided on the latter. Al and I were in Donegal, where the weather had generally matched the darkness of the sites, but this morning was different. The sun was shining and the weather app confidently announced that it would be dry until mid-afternoon. Glenveagh National Park seemed an ideal way to spend a day off and an opportunity for my visibly wilting husband to be cheered by fresh air, blue skies and not a famine village in sight.

At Glenveagh we abandoned the car at the car park and hopped on a courtesy bus to the castle so we could start our walk with a wander around the formal gardens before heading off to the wilderness. The bus was clearly designed to take hordes of primary-school children, as the adults had to sidle sideways down the aisle before squeezing themselves into tiny seats. We were the only non-Americans on the bus. The driver's name was Bernard, but the joker in front of us insisted on calling him Brendan. Everyone (apart from us, and Bernard, presumably) thought this was hilarious. Fortunately, Bernard was good-humoured, and the journey passed with the wag explaining to Bernard why American salmon was much better than the fresh-water salmon in Donegal's rivers and lakes. Bernard begged to differ, and since neither man was prepared to back down, they just got louder and louder on the assumption that the winner would be the last one to lose their hearing. Thankfully, it was a short ride.

Beyond the manicured gardens that surround the castle (really a glorified hunting lodge built in 1867), Glenveagh National Park is a pristine wilderness. The day we were there the early-autumn light flooded the valley and it looked magnificent. We paused for the all-important tea and scone before heading off on a stroll along the edge of Lough Veagh to see the waterfall. On our way back we were hit by soaking squalls of rain, but despite the damp-ness I had begun to relax and could feel the signage stress and tour-guide tinnitus fading away.

Leaving the park, I decided to have a quick look around the visitor centre, just in case. 'In case of what?' said Al. Who knows, but at this point I was finding it hard to walk past any building that might have an exhibition lurking inside. The centre is built in the shape of a doughnut, with wedge-shaped rooms devoted to everything from giant insects to information on the flora of the park. We ambled through it, reading about Molly the Midge and accounts of eagles in the valley, relieved that there was no darkness to be had, but the final room contained a nasty surprise. Glenveagh National Park sits on land formerly owned by 'Black John Adair', a land speculator from County Laois (so much for the idea that all landlords were English) who had created the estate in the mid-1850s. Adair's land agent, Mr Murray, was murdered in 1860 and when, a year after his death, no one had been arrested or charged, Adair evicted all his tenants. Over 240 tenants and their families were thrown out of their homes and off the land. Display boards showed graphic illustrations from contemporary newspapers of poor farmers' homes being demolished. Forty-three families sought shelter at the workhouse in Letterkenny, while others emigrated to Australia. The national park, that glorious stretch of untouched wilderness we had so enjoyed, is a site of forced emigration. There was, it seemed, no escape from the darkness. 'I think I'm going to lie down for a bit,' said Al.

Emigration

અ

'Patience. Fortitude. Long-suffering
In the bruise-coloured dusk of the New Worlds
And all the old songs had nothing to lose.'
— EAVAN BOLAND,
The Emigrant Irish

N

Ulster American Folk Park ■ (8)

■ Ionad Deirbhile
(4)

EPIC: The Irish Emigration Museum ■ (7)

(6) Barack Obama Plaza
■

(2) Bunratty Castle ■

(9)
■ The Great Blasket Centre

■ Dunbrody Famine Ship
(3)

Great Blasket
Island (10)

(5)
Nano Nagle Place ■ ■ Cobh Heritage Centre
(1)

I'm not a fan of organized fun. If there's time scheduled for having fun, you can be sure that I'll be the one sitting in the corner grumbling and looking miserable. So venturing out in the midst of the St Patrick's Day celebrations in Chicago was sure to end badly. But as we were living in the city it seemed churlish to pass up the opportunity to see the greening of the Chicago River. Every year since 1962, the river has been dyed emerald green on the Saturday closest to 17 March. Crowds of up to four hundred thousand people throng the riverfront to watch a boat decant orange powder (made from a secret, and hopefully not toxic, recipe) into the water to celebrate 'St Patty's Day'. We shuffled through the merry crowds, a mix of sober families and people who'd decided to start the day with a Guinness or two. At the east side of Michigan Avenue we stopped to survey the scene. The usually murky river was aglow as the powder turned the water not so much emerald as the green of a traffic light. I leaned over to take a photograph and was accosted by an energetic woman wearing a green T-shirt emblazoned with 'I'm not a Leprechaun, I'm just small' and green

sparkly shamrock-shaped deelyboppers (or zogabongs) bouncing on her head. 'Where's your green?' she yelled, her shamrocks
swaying to and fro. 'Everyone has to wear green!' 'But I am Irish,'
I protested. 'That's not good enough!' she roared, before stumbling off into the crowd.

I've no doubt that the green-clad reveller was one of the thirty-
three million people in the United States who claim Irish ancestry.
Two hundred thousand of them live in Chicago, and it seemed
they were all out on the city streets that day. Many are descended
from the 5.2 million Irish men, women and children who left Ireland for the United States between 1820 and 1920. In the 1860s,
Friedrich Engels was horrified by the scale of Irish emigration
caused by 'conquerors . . . always inventing new . . . methods of
oppression' and mused that if it continued as it was 'for another
thirty years, there will be Irishmen only in America'. Engels
wasn't quite right, but by 1890 there were more first- and second-
generation Irish people in America than there were living on the
island of Ireland. Standing in Chicago that St Patrick's weekend,
I wondered if the further the 'hyphenated Irish' got from the
motherland, both in time and space, the more their sense of what
Irishness entailed became sentimental and commercialized,
reduced to a set of clichéd tropes – wearing green, waving shamrocks, drinking Guinness, yelling 'Top of t' mornin' to ye!' – some
of which bear, at best, a tangential relation to actual Irish culture
and customs (I thought of the bagpipes in the video for House of
Pain's 'Jump Around' and winced). As we wandered around Chicago at its most green and raucous, Barack Obama was giving a
speech at a St Patrick's Day luncheon in Washington, where he

claimed, 'The Irish helped forge the very promise of America: that success is possible if you're willing to work hard for it . . . Through tragedy and triumph, despite bigotry and hostility . . . the Irish created a place for themselves in the American Story.' This is the narrative that Irish-Americans like to tell about themselves, but as I discovered on my tour, it's also the narrative that the Irish prefer – an upbeat and highly selective story in which every emigrant success is championed, and every failure ignored.

We had moved to Chicago on the day of Obama's inauguration, just as Richard J. Daley was heading towards the end of his sixth term as the city's mayor. The Irish had dominated politics and the police in Chicago from the late nineteenth century until the early years of the twenty-first. Indeed, between 1955 and 2011, Richard M. Daley and his son Richard J. Daley had held the position of mayor twelve times between them. The Daleys were from Bridgeport, traditionally a working-class Irish suburb, near the stockyards described by the *Chicago Tribune* in 1889 as being famous 'for smells, for riots, bad whiskey and poor cigars', but by 2009 the Irish were spread through the city and not as clustered in neighbourhoods as they had been. But the strong ties endure. In 1889, 110,000 or 17 per cent of the city's residents were Irish born or had at least one parent who had been born in Ireland; today 7.5 per cent (more than 200,000) of those living in the city claim a family tie to Ireland.

I now live in Liverpool (though the truth is that, if I live anywhere, it's seat 5D of a Boeing 737-800 somewhere over the Irish Sea), but as an emigrant I've never sought out an Irish community – though in Liverpool the Irish are everywhere. There have been days where I've only interacted with Irish people – at work, in shops, on the bus. And walking home through the park I often see people playing hurling and camogie. But despite not joining Irish clubs or societies, I spend my days immersed in all things Irish. I'm paid to think, write and teach about Irish history and the construction of identity and what it has meant over time to be Irish. In that respect, I'm a professional Irishwoman, but not the all-singing, all-dancing variety (very much neither singing nor dancing). I've noticed, when living in the United States or Britain, that people have certain expectations of the Irish. I'm frequently told that we say 'butter' in a charming way, and it never fails to amuse English friends when I talk about putting plates and mugs in 'the press'. It's assumed that everyone knows everyone, that we're all Catholic, and we all like a drink. We're stereotyped as chatty, funny drunks – we're great craic altogether, so we are. And there are kernels of truth in those stereotypes – but they are the inflated, simplified versions, and to satisfy (or confound) them I find involves some degree of performance. I can see how over the years and the generations, when memories of Ireland have faded (or become second, third, fourth hand), the performance is sometimes all that's left. There's something comforting, and indeed in its own way authentic, about the performance, but it's not Irish, or at least not as it would be recognized in Ireland.

Emigration from Ireland didn't begin and end with the Famine. The Irish have always been busy leaving, from St Colmcille in the sixth century to the present day – indeed, statistics from the OECD show that, in 2018, 17 per cent of those born in Ireland were living abroad. Over the centuries many have left reluctantly, involuntarily or in search of new adventures. In the fifteen years before the Famine began, it's estimated that over half a million people left Ireland for North America, while thousands of others took the shorter journey across the Irish Sea to Britain. Emigration peaked during the Famine and the years immediately afterwards, but the exodus of people leaving Ireland continued at a steady pace. Throughout the twentieth century, people left a conservative, repressive society – a state dominated by a Church which allowed no place for them. The country was, as James Joyce put it, 'the old sow that eats her farrow'. The self-imposed exiles of Joyce, Samuel Beckett and Edna O'Brien are well known, but there are thousands of others. But the motives weren't all negative, for hundreds of thousands have left in search of adventure, opportunities, love. Whatever the motive for leaving, the most consistent story of Ireland is the story of emigration. We leave. We don't come back. We romanticize what we left behind. And there is no doubt that many descendants of those who left Ireland don rose-tinted spectacles when they think about 'home'. But those glasses often obscure hard truths.

H. V. Morton once described Cobh as 'the saddest spot in Ireland . . . a wound which Ireland cannot stanch: and from it pours a constant stream of her best and youngest blood'. It was from the pier at Cobh that an estimated three million people emigrated between 1815 and 1970. Some of their stories are told in the old Victorian train station which now houses Cobh Heritage Centre. I took Misha, his sister Matilda, and their cousin Lilo with me. It was lashing rain as I parked the car, and we navigated our way around large, deep puddles to get to the entrance. The centre is on the waterfront at Cobh, and on our right the *Celebrity Eclipse*, a shimmering multi-storey cruise ship, loomed above us. Three hundred and seventeen metres long and thirty metres wide, the *Eclipse* would have towered over those ships that carried thousands of passengers across the Atlantic during the Famine. Even the *Titanic* (two hundred and sixty-nine metres long and twenty-eight metres wide) would have been dwarfed by most modern cruise ships. Nonetheless, the presence of modern liners helps us to imagine what those departures felt like. Some of the passengers on the *Eclipse* leaned over their balconies and waved at us. On our left was the long red-brick wall of the old train station, where a number of notices had been painted, intended to lure us (and no doubt those on the ship) into the heritage centre: 'The *Titanic*', 'Emigration to North America', 'Convict Ships to Australia', 'The *Lusitania*'. Unfortunately, as is no doubt clear by now, most stories to do with the sea, especially those associated with Cobh, tend to focus on hardship, death and disaster. It was no wonder the tourists had stayed in their luxury cabins.

I bought tickets and we each chose a boarding pass. We were given the identity of a passenger on a ship that had sailed from Cobh and, somewhere in the cavernous train station, we would find our alter ego. The kids whipped past the information panels, lined up like a welcoming committee, in search of their story, pausing only to engage with Barry, a lifelike but mechanical ticket inspector decked out in fine livery. Matilda halted as he called her back: 'You – yes, you. Come over here and show me your ticket, please. Come on, I haven't got all day.' She sheepishly returned, brandishing her ticket, and was several sentences into an explanation when she realized that he wasn't real but his recorded messages were perfectly timed to appear as if he was having a conversation.

My boarding pass told me I was Annie Moore, who, aged seventeen, was the first passenger processed through the immigration centre at Ellis Island in New York in January 1892. Coincidentally, the previous evening, while on a ghost tour of Cork, we had stopped at Annie Moore's tiny house on Rowland's Lane. Moore had emigrated with her two younger brothers to join their parents, who had left the year before. As the first passenger through Ellis Island, she was presented with a $10 gold coin which, according to the *New York Times*, 'was the largest sum of money she had ever possessed. She says she will never part with it, but will always keep it as a pleasant memento of the occasion'. Given that her family were living in poverty in a tenement on the Lower East Side, it's highly unlikely that Moore held on to a memento that today would be worth about $275. Moore left home but stayed nearby, and, aged twenty-one, she married Joseph Schayer, a German-American who worked as a fish filleter at Fulton Fish Market. But, for the family, the United States did not prove to be the fabled land of great opportunity. They remained poor, and there was much hardship and grief, for in the first two decades of the twentieth century the couple had eleven children, five of whom died before they were three. Annie Moore herself died of heart failure aged fifty. I wondered if she ever regretted her decision to leave Ireland.

The children's stories weren't much brighter. Matilda was Helen Smith, who was only six years old when she boarded the *Lusitania* in 1915. When the ship sank, both her parents and her brother drowned, but young Helen survived and was brought up by her mother's family in Swansea. Misha was Edward Colley, an engineer from Dublin who was travelling as a first-class passenger. He drowned. I felt sorry for Misha, who had also drowned at the Titanic Experience only a few weeks before. It seemed that Cobh was not inclined to let him live. Lilo was Jeremiah Burke from Glanmire in Cork, a third-class passenger on the *Titanic*. He, too, drowned. Before the disaster struck, he put a note in a small holy-water bottle that his mother had given him and threw it into the ocean. A year later, when the bottle washed up on the Cork coast, Burke's note had acquired an almost unbearable poignancy: 'From *Titanic*. Goodbye all. Burke of Glanmire, Cork'. If there were any upbeat stories, Cobh was keen to keep them from us. Perhaps Morton was right: it really is the saddest spot in Ireland.

I hadn't anticipated writing about Irish slavery in this book. It is true that Viking raiders captured both treasure and people: *The Annals of Ulster* recorded that in 821 Howth 'was plundered by the heathens, and they carried off a great number of women into captivity'. Captives were often ransomed, put to work, or sold at slave markets across Europe. Indeed, Dublin had a large slave market, and a slave chain and collar which is likely to have been made there is on display in the National Museum on Kildare Street. However, by the end of the early medieval period slave trading in Ireland had died out.

On my travels I discovered that very little distinction is made between medieval and modern slavery and, sandwiched between stories of plantation, transportation and indentured servants, I came across references to slaves time and again. In Wicklow Gaol a panel headed 'Slavery' goes on to announce that 'the trade in Irish slaves began during the early 1600s', confusing stories of indentured servants with the transatlantic slave trade. On guided tours, I found that slavery was often mentioned casually, followed by some local tale to illustrate it (tales that all turned out to be anecdotal when I probed further). I don't mind a bit of embellishment, the occasional sweeping statement, but the stories of sixteenth-century and later Irish slavery are fictitious and feed into the myth of false equivalence – our suffering is at least as bad as yours.

Indentured servants were emigrants who had their voyage paid for, in return for which they agreed to work for the person who had paid their fare for a set number of years. Many were cruelly treated and had horrific experiences. Often they worked as labourers or domestic servants, and when the period had elapsed they were free to seek employment elsewhere. For the most part, they stayed in the areas they had emigrated to, and some became landlords themselves. Francis Barrett travelled from Cork to Virginia in 1624 as an indentured servant, and by the time of his death he owned 1,200 acres (and, very likely, a substantial number of slaves) around the Chicohominy River in

Virginia. In the Cromwellian period, orders were issued in parts of Ireland to round up the poor, deport them (often to Barbados) and sell them as indentured servants. But, unlike African slaves, as soon as the period of indenture was over, they were free.

The myth of Irish slavery has been used to diminish the experience of those who really were enslaved – in America and elsewhere – and has often been used by white supremacist groups to promote their racist agendas. I doubt there was such malicious intent behind the stories of Irish slavery I heard on some tours. For the most part, I think it was the result of a lack of knowledge (though that is reprehensible enough), a lack of consideration for the power that words have and the natural tendency to exaggerate or embellish suffering. But whatever the reason, there is no excuse, as the persistence of such myths can have damaging consequences.

Towards the end of my trip around Ireland I got a real and very unwelcome insight into the consequences of inaccurate teaching or storytelling. It happened at Bunratty Castle in County Clare. I'd called into the Folk Park to meet Jean Wallace, the education manager for Shannon Heritage. She took me on a whistle-stop tour of the castle and park so I could tick off my dark-tourism checklist. Murder hole – above the castle entrance – tick; dungeon at the base of a tower – tick; a Cromwellian suit of armour – tick. But the real darkness came at the top of one of the castle towers. Jean and I were looking out over the Raite River (or Owenagarney) when we were interrupted by an American couple. Our conversation began innocuously enough when the tall, thin man, dressed in russet-coloured clothes with matching hair and beard, asked us about the Irish language and where it was spoken. We chatted away agreeably about the Gaeltachts and which one he might visit. Then his wife chipped in. She, short, stout and dressed entirely in denim, was bouncing around in an agitated fashion. Her blonde hair was pulled back in a severe ponytail and she whipped it to and fro as if flicking flies away. She told us she had been taught all about Irish slavery. She 'knew' that 1.7 million Irish people had been forced to become slaves because from 1830 (we never discovered why that year was key) you couldn't import 'African-Americans, hmmm, Africans, hmmm, Blacks, oh, whatever damn term is politically correct these days' into the United States. In particular, she 'knew' that those who left Ireland during the Famine were slaves. Jean and I calmly pointed out that what she 'knew' was incorrect, and we talked about indentured servants and Famine emigration and I suggested she might read work by Liam Hogan and Nini Rodgers. Sadly, it made not a jot of difference. Her husband kept tapping her on the arm in an effort to make her stop, but the diatribe continued. We were wrong. We had been 'brainwashed' (though I'm not sure by whom). It was a surreal and depressing experience, standing on top of a fifteenth-century castle being harangued by an angry, poorly informed woman about Irish history while simultaneously worrying about the fact that at some point we would all have to negotiate the narrow, dark and slippery spiral staircase that led back to the ground floor (I wasn't going to volunteer to go first). I suspect Bunratty got a battering in her TripAdvisor review, but later, as I walked to my car, a middle-aged American couple approached me. They'd witnessed

Jean and me being verbally assailed and wanted to thank us for challenging her assertions. That sort of ignorance has to be tackled, and so it is worrying that some Irish sites and tours continue to perpetuate such a damaging and false story.

Famine emigration is well documented in ships' manifests, newspapers and letters home and it also appeals to many tourists, particularly those with Irish ancestry from North America. Although not all of those ancestors left Ireland during the Famine, the stories of Famine emigration form part of the Irish-American foundation myth. Several museums have full or partial replicas of emigrant ships, but the two most significant are the *Dunbrody*, moored on the River Barrow at New Ross in County Wexford, and the *Jeanie Johnston* on the Liffey in Dublin. I visited the *Dunbrody* with my friend Danielle. Despite the tour being fully booked, an American coach party kindly allowed us to stow away on theirs.

We moved through a darkened corridor lined with information panels and assembled outside a reconstruction of the offices of William Graves and Son, who owned the ship, where we presented our tickets before boarding the *Dunbrody*. We were travelling steerage from New Ross to New York in March 1849 (though the ship we boarded is a replica launched in 2001). The *Dunbrody* was a three-masted barque that regularly traversed the Atlantic from 1845 to 1875 and on board we learned that the ship had been used as a cargo and passenger ship – taking people to the United States and returning with goods. I wondered whether bands had played on the dock as the *Dunbrody* set sail, and I imagined scenes like the one witnessed by Anne Maria and Samuel Hall in Cork in the 1840s:

> Mothers hung upon the necks of their athletic sons, young girls clung to elder sisters, fathers – old white-headed men – fell on their knees . . . Shrieks and prayers, blessings and lamentations mingled in one great cry, until a band . . . struck up 'St Patrick's Day'. 'Bate the brains out of that big drum, or ye'll not stifle the women's cries,' said one of the sailors to the drummer.

From the deck, we negotiated a set of steep, narrow wooden steps that led down into the hull to see the cramped conditions of the steerage passengers. Our tour group numbered fifty, and that felt pretty crowded as we bunched up together on long wooden benches; it was hard to imagine just how unpleasant it would have been with three hundred people crammed below deck for several weeks. To help us, we were joined by two 'passengers' in period costume – Mrs O'Brien and Mrs White – who told us of their experiences. Whole families shared a single bunk, spending weeks in almost complete darkness (it was forbidden to light candles below deck for fear of starting a fire). Access to the deck was restricted to a few minutes a day, if at all. In rough weather, the only people allowed out on deck were those charged with emptying the slop buckets. Mrs White, the first-class passenger, had a slightly more comfortable journey, with proper bunk beds and food supplied by the ship (steerage passengers had to bring their own food – some took hard-tack biscuits and some salted meat, but most survived on oatmeal mixed with hot water). While many may have fallen ill on board, almost all those who travelled on the *Dunbrody* survived.

Captain John Baldwin ensured the *Dunbrody* was well run and regularly cleaned. Those with typhus or cholera were quarantined (in so far as that was possible in the cramped steerage section), but many emigrant ships were overcrowded, often with passengers who were already ill. Some sailed with an insufficient water supply and many made no attempt to separate ill passengers from the healthy. The term 'coffin ship' accurately described many ships which docked at the immigration depot on Grosse Île in Canada. Thousands of dead and dying were taken from them and, during the summer of 1847, 5,400 Irish emigrants were buried on the island.

Avoiding both typhus and cholera, we disembarked and walked through the arrivals hall, hurrying past panels and interactive exhibits detailing life in North America, from the awful situation at Grosse Île through to the Irish-American Hall of Fame, where we read about the Kennedys and other Irish-American success stories, including Maureen O'Hara, Flannery O'Connor and the astronaut Eileen Collins. Until very recently, the Irish-American story was the Irish success story par excellence, reaching its apogee with the election of John F. Kennedy as President in 1960, and it seemed fitting that after they disembarked from the *Dunbrody*, the coach party headed off to visit the Kennedy homestead, a few miles outside New Ross.

Hundreds of thousands continued to leave Ireland after the Famine, in part because of 'chain migration', where family members abroad sent money home to bring other family members over, and in part because many still struggled to survive on the edge, where land was poor, tenant rights were few and the prospect of escaping extreme poverty was limited. One charitable scheme which assisted emigration was the Tuke Fund, which helped 3,330 people from the Poor Law Unions of Belmullet and Newport in County Mayo move to North America in 1883 and 1884 (the fund also assisted thousands of poor in Clifden and Oughterard in County Galway). The fund was named after its founder, James Hack Tuke, a Quaker from York in England. During the Famine he had travelled around Ireland and reported on the distress and suffering he had seen. Tuke maintained a strong interest in Ireland, and in the late 1870s he was horrified by the famine conditions he saw along much of the west coast. In the spring of 1881, he arranged for £1,000 of seed potatoes to be distributed around Belmullet, and when he visited in October he was feted with bonfires lit in his honour. He wrote to his sister, Esther, that he was delighted to see that the potato harvest had been good and that 'in the markets . . . are groups of men and women – the country women with bright kerchiefs on their heads and dark brown or red skirts, and often minus shoes and stockings! Many are selling fish and we hear them chaffering and counting them out in Gaelic.' But he knew this was a temporary fix and was convinced that a measured and well-funded emigration scheme, which assisted families not only to get to America but to establish themselves when they got there, was the only solution to the poverty on the Mullet Peninsula. Tuke's scheme was voluntary, which made it very different from the forced emigration organized during the

Famine by landlords like Major Mahon at Strokestown. Under Tuke's plan, whole families emigrated together and everyone was provided with new clothes for the journey and cash to help them settle in. It was a requirement that at least one person in each family group had to be able to speak English and they had to have a contact in the United States who could assist them when they landed. This was a much more humane resettlement scheme than most of those that had operated during the Famine.

The story of the Tuke Fund is told at Ionad Deirbhile, a heritage centre at Eachléim near Blacksod on the Mullet Peninsula, where, with my friend Agatha, I met Rosemarie Geraghty, who is responsible for the exhibition. Over a large pot of tea and some warm scones she talked about the 'secret famine' on the peninsula in the early 1880s. During her research she had discovered that many of the friends and neighbours who had left maintained their old connections in their new home. In May 1883, Michael and Bridget Caulfield boarded the SS *Manitoban*, bound for Quebec with their seven children. A few weeks later, their neighbours John and Margaret Toole and their eight children left Erris. Both families settled in Miner's Mills, a suburb of Wilkes-Barre, a mining town in Pennsylvania, and the 1900 census lists them as neighbours. Others forged even closer connections: sixteen-year-old Mary Dixon sailed on the SS *Nestorian* with her family in March 1883 and five years later, in Pittsburgh, she married a former neighbour from Mayo. Rosemarie spoke enthusiastically about forging connections with the descendants of the families. Not only was she in touch with many of them, but more than forty families whose ancestors had boarded the fifteen steam ships that sailed out of Blacksod Bay in the 1880s had crossed the Atlantic to visit the centre and the memorial garden beside Blacksod Lighthouse.

But emigration wasn't always born of necessity. Some of it came from conviction. Close to the most northerly tip of the country, at Inishowen Head, you can see the point where Colmcille, the patron saint of emigrants, bade farewell to Ireland and set sail for Iona. Colmcille was one of the first Irish missionaries, and thousands followed in his wake, but that form of emigration rarely appears in museums. Mentions of the Irish nuns, brothers and priests who went on missions in the nineteenth and twentieth centuries are few and fleeting. This may be a reflection of an increasingly secular Irish society. It may be that museums are now reluctant to engage with stories about a largely discredited Church, or because the idea of conversion to a 'better' faith or the imposition of a 'better' culture is, quite rightly, no longer regarded as a good thing.

Thousands of families in Ireland have stories of relatives who left the country to go on the missions. My grandmother and Nancy were neighbours and best friends, and Nana eventually married Nancy's brother Joe. Nana used to tell a story about how, when they were at school, a priest came around recruiting young girls to join an Australian order of nuns – the Sisters of St Joseph of the Sacred Heart. Nancy signed up and, in the summer of 1931, when she was fifteen years old, she boarded a ship bound for Australia. 'We all begged her not to go,' Nana said. 'It was no use. We nearly broke our hearts after her.' Her family never expected to see her again, for Australia was too far away, too expensive and, most importantly, familial contact wasn't encouraged by the order. It was twenty-seven years before Nancy made her first trip back to Ireland, and all of her family stood in the pouring rain in Dún Laoghaire to greet her when she disembarked. 'There was never such rejoicing on this earth as when we met,' Nana recalled.

In my childhood, Aunty Nancy was a very exotic figure because she existed mostly in my imagination. I loved it when we received the blue tissue-paper airmail letters from her. They would be opened with great ceremony and read out loud. Aunty Nancy wrote of a world that was completely alien to me – a world of kangaroos and koalas. I remember the excitement when I met her for the first time – dressed in her pale blue habit, very different from the heavy black habits I associated with nuns. For some unknown reason, I serenaded her with a rousing rendition of the 'Battle Hymn of the Republic' (this from a child who had been asked to mouth the words in the school choir, as my anarchic singing voice was leading everyone astray). Why we learned an American Civil War song in a primary school in North County Dublin in the eighties is still a mystery to me – perhaps it was preparation for emigration. If I'd ended up in the US, I could have integrated seamlessly by bursting into a chorus of 'Glory, glory, Hallelujah!'.

Many years later I worked on the development of the heritage centre at Nano Nagle Place in Cork, which focuses on the life and work of Nano Nagle, the founder of the Presentation Sisters. I was particularly keen to tell the story of the five nuns who travelled from Ireland to San Francisco in the mid-1850s. Dressed in their heavy black serge habits, they travelled west, heading, like so many thousands of others, towards the California gold rush. Indeed, if they had been prospectors, there would doubtless have been a film made by now, but because it's the story of five women it is only found in the pages of academic books and articles. One of the nuns wrote and illustrated her memories of their epic journey – they had crossed the isthmus of Panama on mules, fallen up to their necks in mud, been carried out to tiny boats in Panama (all while wearing their unbearably hot, heavy and increasingly filthy habits) before finally boarding the steamship *Golden Gate*, which, after eleven days on the Pacific, disgorged them at San Francisco. At that time, the city was little more than a rapidly expanding mining outpost filled with thousands of prospectors from all over the world. By the time the nuns arrived, over fifty-two thousand people were living in hastily constructed wooden shacks along streets that eventually formed the backbone of the modern city. The Archbishop of San Francisco was deeply unimpressed by the sisters' arrival. He had asked for Irish nuns, in part because of the huge influx of Irish to the city seeking their fortune, but what Archbishop Alemany wanted was the sort of nun who taught the daughters of the rich in fee-paying schools, not the sort who tended to the poor. But the sisters (no doubt baulking at the idea of the return journey) refused to budge and set up their first convent in a tumbledown shack, before renting two wooden houses, where they taught two hundred children. Despite the lack of welcome from the hierarchy, the nuns settled to their task, and their school and convent thrived and expanded, until the earthquake that devastated San

Francisco in April 1906. The convent and one school were burned by the fire that raged through the city, while other schools were destroyed when the city decided to dynamite buildings in the path of the fire. By the time the embers had cooled, everything the sisters had built up since their arrival in 1854 was gone. They had to start all over again.

In 1995, President Mary Robinson observed that 'Emigration is not just a chronicle of sorrow and regret. It is also a powerful story of contribution and adaptation . . . this great narrative of dispossession and belonging . . . has become, with a certain amount of historic irony, one of the treasures of our society.' It is true that emigration has been 'treasured' (or, at least, engaged with) in Irish culture: in the novels of Anne Enright and Joseph O'Connor, the plays of Brian Friel and John B. Keane, by generations of songwriters, including Shane McGowan, Damien Dempsey and Lisa O'Neill, and in the poetry of Eavan Boland and Bernard O'Donaghue. In politics, however, emigrants tend to be undervalued. Most Irish politicians ignore the diaspora unless a useful soundbite is required; indeed, the state has created Kafkaesque levels of bureaucracy for any emigrant who might actually want to return home. For centuries, Ireland has dispatched millions from its shores; having them return has never really been part of the plan.

But the red carpet is rolled out for the fleeting return of some who have Irish roots – none more so than John F. Kennedy in 1963, Ronald Reagan in 1984 and Barack Obama in 2011. There's a small exhibition about such presidential visits in the most unlikely of locations – the Barack Obama Plaza, a motorway services at Junction 23 on the M7. As the smell of Supermac's burgers wafts up the stairs, I read about young Falmouth Kearney, the son of a shoemaker, who left the nearby village of Moneygall, County Offaly, and headed for Ohio. A hundred and fifty years later, his great-great-great-grandson, President Barack Obama, 'came home to find the apostrophe we lost somewhere along the way'. The undoubted highlight for me (and I suspect every other visitor) is the cardboard cut-out of Barack and Michelle Obama which welcomes everyone who arrives to pay for their petrol. On my visit, I couldn't resist buying a souvenir Barack Obama Plaza pencil and magnet to add to my cabinet.

Barack Obama Plaza is not the only place that tells upbeat emigration stories. EPIC: The Irish Emigration Museum is located on Dublin's Custom House Quay in the basement of an old storehouse built in the early nineteenth century to hold tobacco, tea and spirits. As with so many contemporary museums, my nephew, nieces and I were all given an activity to complete on our trip through emigrant Ireland's greatest hits. In this case, we got a 'passport' to stamp as we travelled through each zone. EPIC's impressively high-tech experience takes visitors from 'Hunger, Work, Community' and 'Conflict' through to 'Discovering and Inventing', 'Leading Change', 'Creating and Designing' and 'Celebration'. The museum feels a bit like a series of PowerPoint slides made to accompany a motivational speech. Even the negative is given a positive spin: one section on notorious Irish immigrants is entitled 'Achieving Infamy'. But my nephew and nieces were engrossed. This was their kind of place.

They dutifully stamped their passports in each section, they slid small discs, as if playing air hockey, across a silky surface in order to learn about the Irish and sporting success, they swiped screens to read letters written by famine emigrants, they decided whether Billy the Kid or Mrs O'Leary was pardoned after their death, and in the section on *Riverdance* Lucy danced a jig across the floor. There is something uplifting about this version of the Irish story but, given the choice, I'd skip some of the glitz and glamour in favour of some actual artefacts and a bit more nuance. In EPIC, I feel like an imposter, that the show isn't for me but for tourists (particularly those from the United States). Certainly, it makes for a flattering tale for all Irish-Americans – their ancestors suffered, their ancestors survived, so they must come from exceptionally tough or driven or talented stock.

The Ulster American Folk Park near Omagh is both museum and theme park, aimed squarely at Americans with Ulster-Scots heritage, and attempts to provide an emigration narrative to rival that of poor Irish Catholics. I was slightly disappointed to find that our tickets didn't involve us taking on the persona of an emigrant – by now I had got used to donning an alter ego for my emigrant travels. First stop was the museum, housed in what felt like a vast hangar where the rain drummed on the corrugated roof. The museum is somewhat sparse and has a dated, eighties feel, full of mannequins in unconvincing dioramas and reproduction portraits of Ulstermen who emigrated and made good. It also has a tendency to gloss over some of the darker aspects of the Ulster-Scots story in America, such as the massacre of twenty Susquehannock (called Conestoga Indians by the settlers) in

Pennsylvania. Over the course of a week around Christmas 1763, more than fifty Presbyterians of Ulster-Scots descent organized a cold-blooded and brutal attack on the Native American population in an attempt to seize control of five hundred acres of land. Of the twenty who were murdered, eight were children. The Paxton Boys (as the attackers were known – a reference to where they lived) were prohibited from taking the land, but none of them was ever charged with murder. At the Folk Park, a panel that mentions the massacre concludes, with staggering understatement, that 'Ulster settlers . . . could be hot-headed.'

The park was originally financed by the Mellon family, whose house is the centrepiece of the park today. The family left Tyrone in 1818 and forged successful careers in the United States, primarily as businessmen and bankers. Thomas Mellon, who was six when he emigrated, became a judge, and his son Andrew's extensive art collection became the basis of the National Gallery of Art in Washington DC. The Mellon homestead is number 9 of nearly forty houses and shops that are scattered across the park's ninety acres of land, taking visitors on a journey from Ulster to a new life in New England. By the end of the journey, visitors are meant to feel as if they have emigrated themselves, and Al and I certainly did, stumbling in the drizzle and pine-needle mulch from one house to another, where we found ourselves occasionally obliged to interact with actors playing characters who had once lived in these very houses, or something like them. Halfway round, we boarded a model ship and headed for Boston. The United States, when we arrived, looked not unlike the land we had left, except that the sun had, conveniently and symbolically, appeared, the grocery store was well stocked, and all the houses were bigger. Just as in EPIC, we were being sold a narrative: emigration as a means of opportunity and self-improvement. All the same, the relentlessness of the message encouraged us to pick up the pace, and we raced through the last four or five homes in record time, returning to the café for some spongy scones while Tiffany rasped, 'I think we're alone now,' from speakers mounted over the till, as if this were how the emigrant story always ended, with a flame-haired woman dancing exuberantly outside a mall.

I'd had enough of positivity and so I headed to Dunquin in Kerry for a few weeks, confident I'd find misery in the home town of Peig Sayers. Unlike many thousands of teenagers forced to read Peig's memoirs as part of the Irish Leaving Certificate syllabus, I had loved the book. I was hooked from the opening line: *Seanabhean is ea mise anois, o bhfuil cos léi san uaigh is an chos eile ar a bruach* ('I am an old woman now with one foot in the grave and the other on its edge') – perhaps her ruminations on death reminded me of my grandmother.

Peig's account of life on the Great Blasket Island is written in an expressive, vernacular Irish and is a treasure trove of details of island life, but successive education ministers failed to appreciate that the average fifteen-year-old has little interest in the daily travails of an old woman living decades before they were born in a part of Ireland most had never visited. As a result, Peig and her book are – somewhat unfairly – remembered by generations of Irish people with the kind of shudder usually reserved for the

Troubles or Brexit: as a terrible aberration inflicted on them against their will, leaving them traumatized and distressed. Certainly, this was the reaction of the woman in front of me in the queue to buy tickets. 'Oh God,' she groaned, 'I didn't know Peig was from the Blaskets. That woman still gives me nightmares.' But I was excited. If *Peig* is the Irish misery memoir par excellence, I thought, surely the Great Blasket Centre will be the dark-tourist site to end all dark-tourist sites, the kind of compendium of suffering and woe in which the very walls emit authentic banshee wails every fifteen minutes (Sundays and Bank Holidays, on the hour). I bought our tickets and stepped inside like a small child entering a funfair.

The Great Blasket Centre is not so much site-specific as site-adjacent. The Office of Public Works (OPW), the government agency running the site, alert both to the environmental impact of encouraging large tour groups to visit the island and the practical difficulties of getting coachloads of tourists across the three-kilometre strait from Dunquin harbour, wisely opted to build an interpretive centre on the mainland, on a headland looking out at the islands. The centre has a curious design. Given its prime position on the cliffs, just a few hundred yards from the mood-setting ruined schoolhouse where *Ryan's Daughter* – that epic tale of misery and betrayal – was filmed, one might have expected a building that stretched lengthways along the cliff, embracing the view with outstretched arms. Instead, like many of the houses in Dunquin, the centre sidles up to the view side on, its windowed ribs offering glimpses of the sea and islands as we were funnelled towards the viewing bench at the end. En route, a series of rooms tells the story of the islands, introducing the famous residents and their works and describing the flora and fauna, the rhythms and rituals, of life on the islands and off. In the first room a fifteen-minute film set the scene, and here was the first disappointment: life, in this account, was pretty good. We saw children dancing on the shore as the water rushes over their feet. We saw men smoking pipes. Much rowing of boats. More dancing. A schoolhouse. Yet more dancing, as if the island was a sort of crystallization of popular Irish tropes, a utopian micro-state-cum-theme park that had somehow managed to filter out all the worst aspects of life while retaining all the diddly-aye charm. At its height, the voiceover informed us, the population of the Great Blasket was 160, but apparently all of them got on. No one ever suffered claustrophobia, or a desperate longing for privacy, or went insane, or drowned, or fell from the vertiginous cliffs at the western end of the island while gathering heather (as happened to one of Peig's sons). The only enemy was the weather. Which left the voiceover somewhat at a loss to account for the island's dwindling population. Still, there is only so much you can cover in an introductory video. I was confident that the stories of desolation were still to come.

Since the islands were made famous by storytellers – most notably Peig Sayers, Tomás Ó Criomhthain and Muiris Ó Suilleabhain – it stands to reason that much of the exhibition is focussed on them. An exhaustive succession of panel texts introduced us to the writers and musicians who lived on the islands – so many of them, in such a small place, that it is almost hard to believe. As we forged a path through several tour groups, Al, who at this stage in my tour had been dragged to more than forty sites, opined that the idea of a small, green, weather-beaten and sparsely populated island with a penchant for overpromoting its writers might be applied to more than just the Great Blasket Island. I decided to let that one go.

The Great Blasket also attracted a succession of anthropologists and linguists, drawn to the island and its people, many of whom they encouraged to write or relate their experiences of life there in celebrated accounts that spread its fame and, in a self-perpetuating loop, drew more linguists and tourists to the island's shores. For all the celebration of storytelling, we didn't hear any of the stories themselves. Peig Sayers, we're told, had a mental treasure trove of three hundred tales, but beyond a very short (and untranslated) audio clip of Peig speaking, we heard none of them. The story I recall most vividly from *Peig* is her account of the 'American wake' held when her best friend Cáit Jim and others emigrated from Dunquin. There was whiskey and dancing, laughter and tears and toasts, for 'We'll never drink another glass together this side of the grave.' Such 'wakes' were held across the country, because until well into the twentieth century emigration was like a death in the family. Many never returned home, and though the arrival of envelopes filled with letters and dollars offered some consolation, there was also enormous sorrow at the departure. When the time came to leave, Peig and other neighbours and friends gathered to walk a little way along the road with those leaving, 'as if it was a funeral procession'. Peig was bereft, for Cáit Jim had been her 'loyal comrade' ever since she was a child. They said a sorrowful farewell and

'Before I had time to wipe the tears from my eyes, they were all swept out of my sight. That was the last time we laid eyes on one another.' The story of Cáit Jim's departure resonated with me because I was a teenager when I read it first, the age when friends are the most important thing in the world, when I couldn't imagine not seeing them every single day, and even if we'd been at school together, in the evening we'd talk for hours on the phone about the latest scandal or romantic misadventure. The idea of one final farewell was impossible to countenance.

Almost all the islanders who survived to adulthood emigrated, most to the United States, some taking a shorter hop to live on the mainland. At the end of the exhibition we heard an account of the evacuation of the Great Blasket in the winter of 1953. By then, the population had dwindled so severely that the islanders could man only two boats, and sent a desperate telegram to the Taoiseach, Éamon de Valera, warning that they were on the verge of starvation. But the centre can't resist closing on a positive note. The final room introduces descendants of the islanders now living in the US, suggesting, like EPIC, the Ulster American Folk Park and the *Dunbrody* famine ship, that the Irish story does end happily, if only abroad. Yet, for its youth, the Great Blasket Island was a place to escape from, a prison ship moored permanently offshore. In that sense, it really is the Irish story in microcosm.

I wrote sections of this book sitting at a kitchen table in Dunquin in between staring out across the fields as An Fear Marbh drifted in and out of sight through the mist. It's the island beside the Great Blasket and its official name is Inis Tuaisceart (North

Island), but it's known by many as An Fear Marbh ('the Dead Man') because the outline, visible over the yellow irises and purple foxgloves that filled the garden, is very clearly that of the body of a man laid out in preparation for his wake.

Every morning in search of inspiration I walked up to Peig's grave, marked by a simple limestone headstone in the small cemetery that perches above Dunquin harbour. Three kilometres offshore lies the Great Blasket and, from Peig's grave, I could see its white cottages glinting in the sunlight above the long sandy beach. It looked beguiling from a distance, and on a warm late-spring day we took a boat across the strait with a handful of other visitors. The beach was deserted save for at least one hundred seals and their pups playing in the shallows. And as we did a circuit of the island we could see basking sharks and dolphins following the fishing boats in the sea below. A myriad of sea birds dotted the sky, their calls caught by a blustery wind. It could not have been more glorious. And yet, those who lived on the island were in constant fear of death, by drowning or falling from the steep cliffs, and at best were eking out a subsistence life on the island. They brought up their children for export. It's no wonder that the arrival of the post on the island several times a week was the highlight.

But the post didn't always bring joy. On display in the Blasket Centre is a transcript of a letter written in 1889 by Jim Boland, father of Peig's friend Cáit Jim, to his brother in Springfield, Massachusetts. His brother had left Ireland some fifteen years earlier and had never written home. There is something plaintive and heartbreaking about this letter, and Jim's sense of loss and hurt is clear:

'Had I known where you were this long time past I would certainly have sent you many a letter . . . But you knew all the time
where I was and I am surprised that you never wrote to me. One
would fancy that two brothers, situated on both sides of the
Atlantic Ocean, would often communicate with each other and
inform about the state of their mutual affairs. Still you have not
written one line to me for the last fifteen years . . . Answer
this . . . May God bless ye . . . is the prayer of your fond brother.'

There is nothing to indicate if Jim Boland ever got a reply
from his brother. I wonder what happened to Cáit Jim's uncle.
Perhaps he had embraced his new life and discarded the old.
Professor Kerby Miller has observed, 'given endemic rural jealousy, every family's pretence that its own children were
prospering overseas encouraged the common notion that America was so rich that anyone could prosper there'. But maybe he
hadn't prospered, or if he had, maybe he didn't want to send
remittances home – those dollar bills that helped sustain many
Irish families, both keeping food on the table and helping to pay
for the passage, which facilitated chain migration and allowed
generations of young adults to leave an Ireland that had little to
offer them. Perhaps Jim's brother just never liked him. I find
such unfinished stories deeply moving.

On the afternoon of Sunday, 15 August 1999, I pushed open the
door of a small, decrepit pub in the East End of London. Clare
were playing Kilkenny in the All-Ireland hurling semi-final, and
I was determined to watch it. These were the glory days of Clare

hurling. In 1995, Clare had won the All-Ireland, finally putting to rest the curse of Biddy Early that had been placed on the team in 1932. Clare won again in 1997, and I fancied their chances in 1999. Despite never living in Clare, I was a fervent supporter of Clare hurling. I had to be. My grandmother was obsessed, and there was no way I could have lived with her without being able to recite the names of the Clare hurlers. The Lohan brothers, Anthony Daly, 'Sparrow' O'Loughlin, Nana loved them all, but her real affection was for Jamesie O'Connor, the slim, blond talisman. There was nothing he couldn't do. I'm convinced that, given the chance, in the triptych of the Pope, de Valera and John F. Kennedy that hung in her sitting room, she'd have replaced Pope John XXIII with O'Connor.

I arrived at the pub about fifteen minutes before throw-in. It was a small L-shaped bar with a handful of men sitting at the counter and another handful at the small round tables that lined the wall on my right. A television was propped up over the door that led to the toilet. I sat on a stool, facing the television. The middle-aged woman behind the bar walked over to me and I ordered a glass of Guinness. 'I'm afraid I can't serve you,' she said. 'What?' 'I'm afraid I can't serve you,' she repeated. 'I just want to watch the match,' I said (perhaps she thought I was under eighteen . . .). The woman shook her head. 'I can't serve women.' She named another pub showing the match – a bus-ride away. I said it was illegal not to serve women and she said she'd have a chat with the men at the bar. I waited, quietly seething. A few minutes later she returned. 'You can stay if they can leave their caps on,' she said. I looked around and realized that every single man in the pub was wearing a tweed jacket and a matching cap. They thought if a woman came in, they'd have to take their caps off, and they weren't comfortable doing that (apparently, the woman behind the bar didn't count). I said they could do what they liked with their caps and I got my Guinness and settled down to watch the match. When it became apparent that I knew what a side-line cut and a sixty-five were, the slight tension in the pub disappeared, and as the game progressed I got chatting to the men. They had all come to England in the fifties, part of a huge wave of post-war immigrants that saw three-quarters of a million Irishmen and women move to Britain. Most of the men in the pub had moved around for decades, following building jobs, before finally settling in London. None of them had married. None of them thought of London as home. Home was Moyasta and Galmoy and Cloonfad and other towns and villages, but they hadn't been back in years, for almost everyone they knew there was gone or buried. Every Sunday after Mass they assembled in this pub to watch a match (GAA in the summer, soccer in the winter – the GAA ban on watching foreign games hadn't bothered them one whit) and sat in silence or reminisced. They talked about how the GAA clubs in England had once held the community together but now, with Ireland booming and more exotic locations on offer, the clubs in London were struggling to field teams. Their world was dying and no one wanted to hear about it.

Fifteen minutes before the end of the match D. J. Carey stepped up and fired the sliotar past Davy Fitzgerald. Kilkenny were a goal up and went on to win by four points. As the final whistle blew and I sat, dejected, on my stool, a ruby-coloured glass of beer was pushed down the bar towards me. I looked up and one of the men at the top of the bar caught my eye and winked. 'It's a glass of Kilkenny. It'll do you good.' And after that, every Sunday I was in London I popped in for a chat with my elderly friends and listened to their stories. I'm sure they're all dead now. As I sought out stories of emigration in Irish museums, their stories barely merited a mention. Perhaps the stories of Cáit Jim, her uncle, my Aunty Nancy and my flat-capped friends are told where they landed and settled. But that's another book.

Signposts

Brown road signs used to excite me. They're like Forrest Gump's box of chocolates: you never know what you'll find when you follow one. Perhaps nothing as exciting as a pot of gold at the end of a rainbow, but at least a megalithic tomb, a ruined castle, a monastic site or an old graveyard. That was more than enough for me. I know it's not exactly daredevil, but there was always a thrill of anticipation as I barrelled along the ever-narrowing roads with grass tickling the underside of the car. But now that thrill is dulled, for these days it seems anyone can put up a brown sign and, as I've completed laps of the country, I've noticed brown signs trying to lure people towards B&Bs, pottery shops and organic-coffee outlets. At every crossroads, fistfuls of them point in all directions like a demented palm tree. At one junction near Glengarriff in County Cork I spotted twenty-six directional signs – fourteen of them pointing me towards a farm, a B&B, a cable car, a hotel, some gardens, a ferry, a cycle route, an arts centre and a caravan park. In the middle of Clifden, nine brown signs clinging to a pole direct visitors to hotels, restaurants and a golf club. It's not quite as bad as the situation in the UK (where one brown sign in Plymouth points towards a drive-through McDonald's), but it's not far off. And yet, despite the persistence of rogue brown signs which occasionally mean I end up at a motorhome-hire outlet when I was hoping for a dolmen, I still regularly turn off the main road to go on a mini-adventure.

I thought that some dramatic policy change might explain the proliferation of the trusty brown sign, so I decided to investigate. It turns out that there are rules – lots of rules. The National Road Authority (NRA) has a twenty-page document outlining its 'Policy on the Provision of Tourist and Leisure Signage on National Roads'. Those twenty pages deal only with roads where the speed limit is over fifty kilometres per hour. Roads with speed limits of less than that are left to the management of the local council. (I thought the NRA's document was remarkably thorough until I discovered the equivalent policy document for Northern Ireland runs to fifty-two pages.)

I found the NRA document bizarrely fascinating. Did you know that on a motorway or a dual carriageway you can only have tourism signage for attractions that get more than fifty thousand visitors a year? On other national primary roads, signs are allowed for attractions that get more than fifteen thousand a year, while on national secondary roads a site only needs to get a paltry seven thousand visitors to be rewarded with a brown sign. So it seems big, successful attractions are allowed additional advertising, while smaller sites that would benefit from the attention get none. However, there is a way around this. If a site can claim 'national importance', it can have a brown sign no matter how many visitors arrive (though the NRA document, for all its thoroughness, gives no indication about what might constitute 'national importance'). I wonder how visitor numbers are calculated for sites that have no visitor centre – does someone inspect the ruin and estimate figures by how trampled the grass is? There is a section on 'continuity signage', which makes me wonder when signage split, and the document ends with a 'sign provision matrix', which is even less exciting than you'd think.

You would imagine that every national monument is of national importance, even if it gets only a handful of visitors, and many of them do indeed have brown signs guiding the way. I follow them with a devotion that reflects the triumph of hope over expectation. One afternoon, while driving from Greencastle to Malin Head in Donegal, I got distracted by a brown sign luring me to Clonca Church (one of over 750 national monuments in Ireland). I parked the car in a little indentation between the road and the hedge, hopped over a stile and strode across the field towards the ruined church. The information panel told me very little other than there were two crosses – one standing, the other lying down – and that inside the church was the tombstone of Magnus MacOrristin.

I went inside the roofless church to find the late-medieval tombstone, and it's one of the most unusual I've seen. A stylized cross carved in the centre has swirls at the base, as if the cross has grown roots. At the top, two serpents' heads are entwined in the shape of a heart, and along the tombstone's right side, next to a sword, is a carving of what looks remarkably like a hurl and sliotar. This may well be the first visual representation of hurling (ironic that it's in Donegal, not a county known for its hurling prowess today). The origins of the tombstone are unclear. It's possible that MacOrristin was a Scot and the carving could represent shinty or golf. But we do know that hurling (or a version of it) was being played in Ireland, both in mythology and reality. Cúchulainn, the Irish hero of the Ulster Cycle (a collection of Irish myths and sagas), was a great warrior, famed for his bravery and strength but also skilled with a stick and ball. It was such proficiency that saved his life when he was attacked by a vicious dog. He picked up his hurl and hit the sliotar with such force and accuracy that he killed the animal. In 1366 a statute passed in Kilkenny banned 'horlinge' in the hope that men would 'throw lances' and 'draw bows' instead (presumably to make their arms longer). But whether MacOrristin's tombstone shows a hurl, a shinty stick or a golf club doesn't really matter. None of these games existed in their modern format during his lifetime. What is truly remarkable is the fact that there is a late-medieval carving representing sport, a rare reminder of daily life and recreation amid the symbols of religion and war.

Incarceration

&

'Every prison that men build
Is built with bricks of shame.'
— OSCAR WILDE,
The Ballad of Reading Gaol

N

1 St Connell's Museum

8 Crumlin Road Gaol

Downpatrick County Museum ■

Roscommon Gaol ■

7 Mountjoy Prison ■

9 Kilmainham Gaol

6 Nenagh
Heritage Centre

3 Carlow
County Museum

Wicklow Gaol ■

4 Tarbert Bridewell

Tipperary
County Museum ■

Reginald's Tower ■

2

5 Elizabeth Fort

Youghal Clock
Gate Tower

Cork City Gaol ■

10 Spike Island

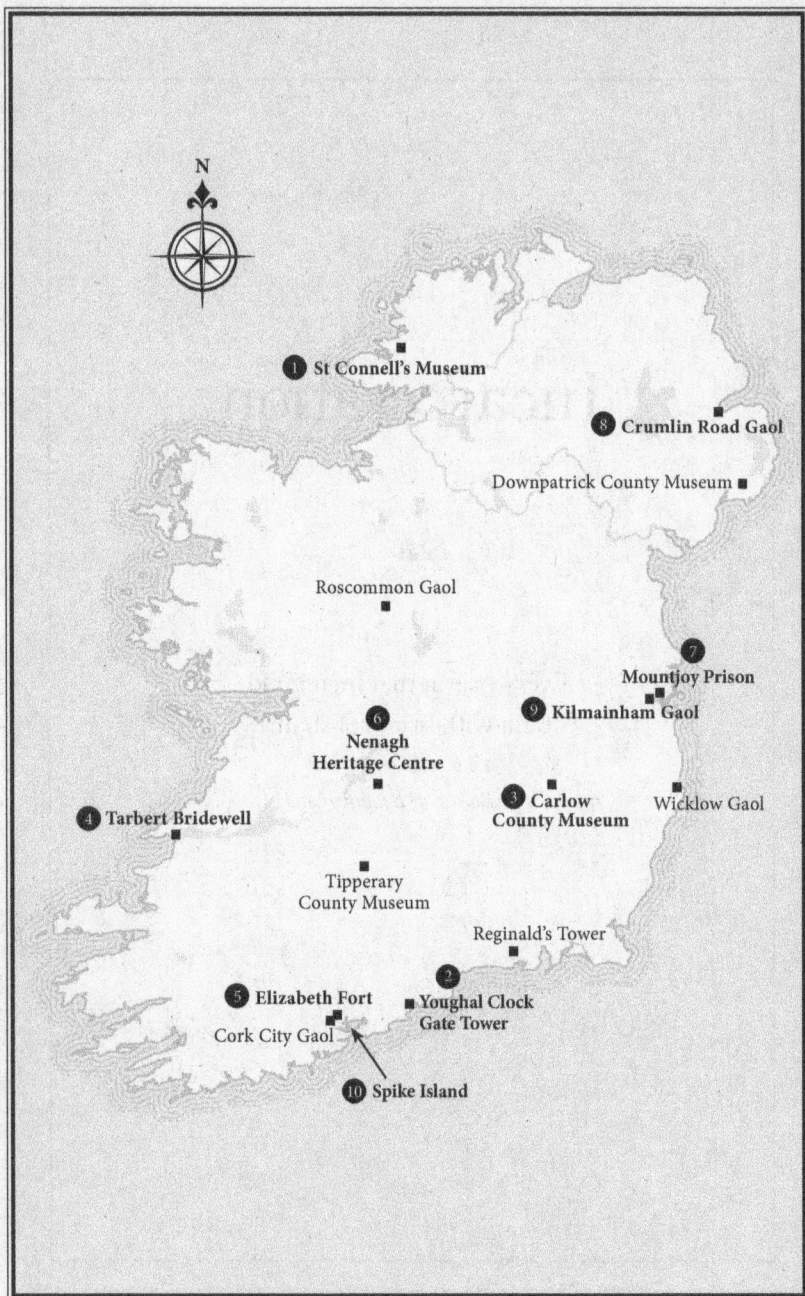

The road from Dunfanaghy to Glenties in Donegal is a beautiful stretch, or so I've been told. But as I drove along it one September day the rain poured down and the wind buffeted the car. As I neared Gweedore I glanced left, hoping to admire the distinctive cone-shape peak of Errigal rising high above the Derryveagh Mountains. But there was no mountain to see. Not even a hint. Donegal was wrapped in cold, wet, impenetrable greyness.

Glenties seemed deserted. Anyone with any sense was at home sitting around a fire. Unsurprisingly, Al and I were the only visitors to St Connell's Museum. Tucked round the corner from the main street, the museum sits beside the courthouse, a substantial five-bay two-storey building completed in 1843 which, at the time, must have seemed incongruous in a small rural village.

The museum's collection is a testament to the charms of haphazard accumulation and display, the three floors offering, in no particular order, part of the fuselage of a Second World War bomber that crashed in the Bluestack Mountains, an exhibition of dramatist Brian Friel and writer Patrick McGill memorabilia, the switchboard from the local post office, a room dedicated to disused railways and a corner detailing the fluctuating fortunes of Glenties in the Tidy Towns competition. But I was there for the cells: a network of holding cells in the basement of the court-house. By now my sense of mission was ruthlessly honed, and I tried not to tap my foot impatiently as a friendly member of staff led us slowly downstairs, stopping to deliver a long disquisition about the aunts of Brian Friel who provided the inspiration for *Dancing at Lughnasa*, and out through the fire exit into a court-yard scattered with famine pots. She pointed us towards the basement of the courthouse and left us to explore the cells. We did a quick inspection, staying long enough to soak up the melancholy in the small damp rooms with their peeling paint, tiny windows and heavy steel doors before retracing our steps across the yard. It was at this point that we realized our attempt to immerse ourselves in a site of incarceration had acquired a whole new depth. Our guide had let the fire door swing shut behind her. Since it didn't open from the outside, we were now trapped in the yard between the cells and the museum.

Every year, over 750,000 people in Ireland go to jail, at least for a few hours. One of the reasons they are so popular is that they possess something no purpose-built museum can offer: authenticity. Prisons are particularly powerful sites of immersion, since the suffering that took place there – from the daily grind of prison life to hangings and executions – was so acute. Walking those dank corridors, hearing your own footsteps echo on the flagstones, standing in the claustrophobic cells, offers a macabre kind of thrill. At least until you find you can't leave.

Up until the late eighteenth century (and in some jails far beyond that), prison was about punishment, not reform. Jailers were very badly paid (and often maintenance for prisoners came out of their allowance so there was little incentive for a governor to try to improve the prisoners' lot). I got an insight into the hardships they endured when I took Misha, Matilda and Lilo to visit the Clock Gate Tower in Youghal. The tower, built in 1777, operated as a prison, accommodation for the jailer, and a clock and bell tower. Our guide, Shirley, was informative and enthusiastic as she told us about the exotic spices and furs that were traded through Youghal's port before moving on to the tower's history as a jail. Out of consideration for the young children on our tour, she skipped over some of the more gruesome tales, much to the disappointment of Matilda, who is every bit as fascinated by the macabre as her brother. But Shirley spotted her downcast face and promised to fill her in later about the two United Irishmen who were hanged from the windows of the tower in 1798. The jail, she went on, occupied the second floor – a room the width of the street. Men, women and children were locked up there together – a month for making poitín, three months for stealing a tablecloth, indefinitely for a debt. Sarah Roche, for example, owed her brother money, and he had her imprisoned until she agreed to marry one of his friends. As we stood in the jail, huddled in our coats – it was July, but a cold, wet, miserable day – Shirley told us we were having a much cosier experience than the prisoners had enjoyed, for back then there was no glass in the windows, partly because of a belief that fresh air dispersed any miasma, or foul vapours, that spread disease. Some also thought it would cleanse the prisoners' souls. Besides (and this was possibly the key reason), glass was expensive and the comfort of prisoners was far from a priority. But the key deprivation was even worse: there was no food. Every day the prisoners lowered a bucket from the window to the street below. Sometimes passers-by put food into the bucket, sometimes they didn't. If the bucket was hauled up empty, there were no meals that day.

I wandered over to a corner of the room to read the list of equipment required for the jail – it included two pairs of blankets, four rugs, a slop bucket and a brush – and turned back to see Matilda sprawled across the cell mattress. She had shackles binding both her wrists, shackles that were attached by a chain to an iron ring embedded in the floor. She was grinning as Misha took a photo, but with a jolt I realized that, at ten years old, Matilda was old enough to have been a veteran of the prison system, had she been alive in the eighteenth or nineteenth century.

Youghal was by no means atypical. Until well into the nineteenth century, jails were fairly rudimentary places built purely for the purpose of confinement and punishment. While one newspaper reported in 1819 that at Reginald's Tower in Waterford women prisoners 'were in the habit of indulging in jigs, reels and country dances to while away the tedious hours', conditions in most prisons were not designed to inspire dancing. The prison system was notoriously corrupt, with warders routinely offering favourable treatment to those who could afford to bribe them, and brutal punishment to those who could not. In the late 1780s, Jeremiah Fitzpatrick, the first Inspector General of Prisons, quoted the eponymous heroine of Daniel Defoe's *Moll Flanders* when he described a visit to Wicklow Gaol: 'That hellish noise, the roaring, swearing and clamour, the stench and nastiness . . . join'd together to make the place seem an emblem of hell itself and a kind of entrance into it.' But by the end of the century new prisons began to reflect the influence of reformers like Jeremy Bentham, John Howard and Elizabeth Fry. The Regulation of Prison Bill, passed in 1786, ensured that prisons were formally inspected. Wages rose for prison staff, though bribery, corruption and brutality remained rife.

The change from a system almost exclusively about punishment to one which encouraged criminals to reform also sparked a revolution in prison design. Crumlin Road Gaol was designed by Sir Charles Lanyon, who was responsible for many of Belfast's Victorian landmarks, including Queen's University, Belfast Castle and the Custom House. It opened in 1846 and, like Mountjoy Prison in Dublin, was modelled on Pentonville Prison in London, with four wings radiating from a central hall in a fan shape. These prisons were designed for the 'separate system', where each cell held only one prisoner. Inmates were, for the most part, prevented from communicating with, or even seeing, each other. This system was intended to force prisoners to reflect on their crime and to prevent minor criminals from being influenced by hardened ones. The separate system was part of a wider approach known as Walter Crofton's 'Irish System', which was based around three phases of imprisonment. The first was a period of solitary confinement at purpose-built jails such as Mountjoy or Crumlin Road (though overcrowding often meant that the separate system was more aspirational than practised). During the second phase of confinement, prisoners were moved to jails such as the one on Spike Island off Cobh, or Kilmainham Gaol, where they could gain credits or 'marks' for work carried out and good behaviour. In the final phase, prisoners were sent to a less strict 'intermediate prison' such as the one at Fort Camden (now Camden Fort Meagher) at Crosshaven, where they could earn a 'ticket of leave', an early form of parole. The new system was a marked improvement on the one that preceded it, with regular inspections and guidelines governing diet, bedding, clothing and education, but it was still a punishment system founded on hard work, strict discipline and religious instruction.

Prison reform is most clearly reflected in the East Wing of Kilmainham Gaol. Designed by John McGurdy, it opened in 1861 and reform ideas were built into the design. Unlike earlier prisons, the East Wing is horseshoe-shaped and based on Jeremy Bentham's panopticon design. There were two key elements at play – observation and light. A warder standing in the centre of the wing could see every one of the ninety-six cells. There were spyholes in every door, while thin strips of fabric that ran along the corridor enabled warders to approach a cell door silently and peer in (in Cork City Gaol, there was no fabric strip, but prisoners and warders wore felt overshoes to deaden the sound of their footsteps). All cell doors opened into the vaulted space, illuminated during the day by light streaming in through the huge glass canopy overhead. In every cell there was a window located above head height, drawing the prisoner's gaze towards the heavens. The religious symbolism was not coincidental. Prisons came with chapels and chaplains, and reformers hoped that, by ensuring that prayer and reflection were part of daily life, criminals would turn to God and, once released, not darken the door of a jail again.

As I visited more and more museums and heritage sites, it became increasingly apparent that far more stories are told about rich and powerful individuals than about poor people. In some ways, this is inevitable. The rich leave more behind – in terms of both written material and artefacts. Even at former prison sites where the vast majority of those imprisoned were the very poorest of society – often jailed for very minor offences – we hear disproportionately more about the political prisoner, the prisoner with a cause, than we do about the pickpocket, the prostitute or the pauper.

Almost every jail in Ireland housed political prisoners at some point, but they never accounted for a significant percentage of the prison population. Kilmainham Gaol held far more political prisoners than any other, including Robert Emmet – the leader of the 1803 Rebellion – Charles Stewart Parnell and Constance Markievicz, but even there less than 15 per cent of the prisoners were classed as political (and most of those were held in the period 1916–23). Nevertheless, tales of political prisoners dominate every prison museum in the country, even when their time in the prison was fleeting. In 1848, the Young Irelander John Mitchel was found guilty of treason felony and sentenced to fourteen years' transportation. Mitchel spent a mere four nights on Spike Island before being transported, but the brevity of his stay did not prevent the fort on the island from being named after him. The reason for this focus on political prisoners is twofold: firstly, some of them have had considerable impact on Irish history (Markievicz, for example, was commander of the garrison at St Stephen's Green during the Easter Rising and the first woman elected to Westminster in 1918) and telling their stories is one way of attracting visitors, and secondly, because the information exists – we know what led to their arrest, about their incarceration and what happened to them after their release (if they made it out alive).

The treatment of political prisoners varied and, as with the regular prison population, the variation was often based on class. A visitor to Kilmainham Gaol in 1798 described it as more like a 'tavern than anything else – card parties, dancing and singing'. Mitchel wrote about his time on Spike Island in his *Jail Journal*, describing the island as 'a rueful looking place', but he was well treated during his brief stay:

Mr Grace [the prison governor] . . . said . . . that he had a letter from [Dublin] Castle directing him to treat me quite differently from a 'common convict', to let me wear my own clothes, not to put me in irons . . . My wardrobe too is somewhat scanty for a 'Gentleman' seeing that they brought me away . . . in an old brown summer coat, old shoes and a glazed cap . . . Mr Grace has kindly taken the trouble of procuring for me at Cove [Cobh] a few changes of linen and other small indispensables.

In addition to new clothes, the governor lent him some books, and Mitchel whiled away the hours reading Shakespeare's *Merry Wives of Windsor.* Given how overcrowded every prison in Ireland was in 1848, it's clear that Mitchel was afforded exceptional treatment.

In 1882, when Charles Stewart Parnell, a Member of Parliament and Leader of the Irish Parliamentary Party, was held in Kilmainham along with others associated with the land movement, it seemed that political prisoners, or those of a certain standing, were still being treated well:

> the great central hall ... was noisy – and very cheerfully so – from morning until night. A long table down the centre ... was littered with the newspapers, magazines, and books of the day; draught-boards, chess-boards, backgammon-boards, and packs of cards. The same table at the dinner-hour bore a cloth of snowy linen, was decorated with fruit, flowers and cut glass, and upheld a weight of excellent hot dishes and wines.

Only a select few were afforded this treatment, for Parnell noted that during his time in the prison many of the prisoners had a 'semi-starved aspect' and 'seemed to be utterly dejected and weak'.

Some political prisoners were singled out for unwelcome attention. Robert Kelly, a Fenian, was sentenced to fifteen years' penal servitude on Spike Island for wounding a policeman. Because many of the warders believed he had also killed Thomas Talbot, an RIC head constable (although he was found not guilty of this), he 'never drove a nail or sawed a piece of wood but his special warder was at his elbow . . . Never out of the right hand of . . . these worthies was a heavily lead-laden baton, nineteen inches long, and weighing over 1lb.' Kelly was kept in the prison block and noted that 'Most of the men here are chained and many of them suffer from terrible diseases such as scrofula, ulcer and skin affections [sic]. We were obliged to drink out of the vessels used by these men; this was most revolting.'

In 1919, during the War of Independence, Constance Markievicz was held in Cork City Gaol after she was found guilty of taking part in an unlawful assembly and using words 'to inflame the passion of the people against the police'. After her release, she commented that the jail was 'the most comfortable I have been in yet', and she had had plenty of experience, having spent time in Kilmainham, Mountjoy, Aylesbury and Holloway prisons following the 1916 Rising. Several years later, during the Civil War, a young Michael O'Donovan (later better known as the writer Frank O'Connor) was less complimentary, describing his brief time in the jail as a 'nightmare', a place overcrowded and 'seething with vermin'.

Highlighting famous prisoners is certainly one way to attract visitors, though, time and again when I spoke to people who had just completed tours at prison sites, I heard how the stories that resonated with them were the stories of the ordinary prisoner, not the political ones. And almost all the visitors I spoke with instinctively identified with the prisoners rather than the warders, though often they couldn't articulate why. I think there are a number of reasons for this. Prison sites tend to focus on the 'ordinary decent criminal' and political prisoners, rather than on rapists or murderers. The stories told at prison sites often emphasize what we would now regard as injustice – stories of imprisoned children, of torture, of transportation, of miscarriages of justice. Victims of crime are rarely mentioned; instead, the inmates themselves are often presented as victims, both of poverty and of the system. And the passage of time romanticizes the criminal. Another reason is that most of the prisons now open to the public operated primarily when Ireland was under the control of Britain, and visitors tend to see the inmates as victims of British oppression, targeted not because they were guilty of a crime but simply because they were Irish. But the prison records for England, Wales and Scotland show great similarity with Irish records. The poor, wherever they lived, were imprisoned or transported. According to Cal McCarthy in *The Wreck of the Neva*, Ireland accounted for about a third of the combined population of England, Ireland, Scotland and Wales but only about a quarter of those transported, which suggests that it was the industrial poor of England who were targeted, rather than the (largely agricultural) poor of Ireland. It's not nationality that divides us but wealth.

It seems safe to assume that most prisoners were guilty of the crime they were accused of (though we may argue both about what ought to have been considered a crime, and what punishment might have been proportionate). The most common crime that landed people behind bars was debt, but the prison registers list many others, including assault, shoplifting, pickpocketing, rape, highway robbery, murder, bigamy, cattle-stealing, mail robbery, counterfeiting coins, illicit distilling and vagrancy. And prisoners came from all walks of life. The prison registers for Kilmainham Gaol between 1830 and 1880 list over two hundred different job descriptions, including acrobat, bone-gatherer, umbrella maker, police constable, gambler, rat-catcher and gentleman.

What we know about prisoners is often restricted to the information contained in the prison register – name, details of the crime and sentence, height, religion, occupation, physical description. Many prisoners were repeat offenders, often appearing in registers under different names, though few were as inventive as Patrick Mahoney, a largely unsuccessful burglar who, over the course of fourteen years (1885–98), adopted fourteen different aliases, including Titus Oates, Septimus Snooks and Baby Nuggins. Before the advent of photography and mugshots, notes were often taken of a prisoner's physical characteristics (in part to try to prevent different aliases being used). In 1827, John Drake, a tailor, was sentenced to seven years' transportation for the theft of a horse. His record states that he had three horse-bite marks on his chest, suggesting that he wasn't a very accomplished thief.

The records offer us glimpses into another world, a strange, cruel and alien one where punishments were brutal and often disproportionate, where a prisoner found guilty of theft could be given the same sentence as someone found guilty of murder. In 1839, eight-year-old Alicia Kelly was sentenced to five months' hard labour for stealing a cloak, and in 1854 John Rooney, who was fourteen, was locked up for three days for shooting marbles and annoying passengers on a train, while in

1864, thirty-three-year-old Bridget Dwyer spent a month in prison for using blasphemous language. And there are thousands of Alicias, Johns and Bridgets who are only remembered today because of a sentence or two in a ledger.

Occasionally more detail survives – particularly when it's an unusual or heinous crime which attracted newspaper attention. In June 1852, a murder made headline news. The painter William Bourke Kirwan and his wife, Maria Louise, took a trip to Ireland's Eye, an island just off Howth, County Dublin. Later that day Maria Louise was found dead in a rock pool. Kirwan claimed his wife had wandered off while he was painting and drowned, but he was arrested and charged with her murder. The press lapped up the story, not least because it was revealed that while he and his wife were living on Merrion Street in Dublin, he had maintained a second home in Sandymount with Teresa Kenny (who also called herself Mrs Kirwan) and their eight children. At the trial Kirwan was found guilty and condemned to death, but his sentence was later commuted and he was sent to Spike Island to await transportation to Bermuda. While they awaited transportation, many of the convicts worked outside the prison walls, but Kirwan, as a nineteenth-century celebrity, was kept within the fort, as there was 'too much curiosity about him'. His story is told on Spike Island, where his paintings once adorned the chapel, though no trace of them remains today.

In a corner of Tipperary County Museum in Clonmel there's a replica of a large luggage trunk once owned by Vere Thomas St Leger Goold, who was born in Clonmel in 1853. Goold, a charismatic confidence trickster, was also a talented athlete. In 1879 he reached the Wimbledon tennis final, which, press reports suggested, he played with 'a roaring hangover', which may explain why he lost. He soon swapped the tennis circuit for a louche lifestyle, travelling around Europe, relying on his charm to find wealthy benefactors to fund his lavish tastes. He married a French dressmaker, Marie Violet Girondin, and by 1907, when they arrived in Monte Carlo, determined to make their fortune at the roulette table, they were calling themselves Lord and Lady Goold. Luck was not on their side and, out of funds and friends, they fled Monaco. Their flight came to a halt at Marseille, when the station master noticed blood oozing from the base of their trunk. Police forced it open and found the naked bloody torso of a woman – you can see a horrifyingly realistic re-creation of the sight that greeted police that day at the museum. The Goolds were staying at a nearby hotel, and a search of their room revealed a case containing limbs and a head, while stuffed inside 'Lady Goold's' handbag was a large quantity of valuable jewellery. Police quickly ascertained that the body was that of Emma Levin, a wealthy Swedish widow who had been befriended by the Goolds. When she refused to give them any more money, they killed her, carved up her body, pocketed her jewellery and travelled to Marseille, where they intended to dispose of the cases before boarding a ship to England. After a high-profile trial, the Goolds were found guilty of murder – Marie Goold was sentenced to death (later commuted to life imprisonment) and Vere Goold sentenced to a lifetime of hard labour on Devil's Island in French Guinea.

On a lighter note, one of my favourite stories is that of James

Gray, a twenty-eight-year-old thief from Manchester who became known as Jack-in-the-Box. Gray built a large wooden box with hinges and springs that enabled him to open it from the inside. He would address the box to someone in Cork, Belfast or Limerick, always writing 'THIS SIDE UP – WITH CARE' on the outside, and hide inside it while an accomplice took the box to a railway station or packing office. Once the box was safely stored in the luggage compartment of the train, Gray would climb out and rifle through all the baggage, stealing anything of value. When he'd finished, he climbed back inside the box and continued his journey. His thefts went undetected until one of his housemates noticed a piece in a newspaper offering a reward for the recovery of some silk shawls which he had seen in Gray's room. He contacted the police and Gray was charged, found guilty and sentenced to four years' penal servitude in 1856.

Deciding what stories are told at a site is a complex and difficult task. I have some experience of this, having been the historical consultant for the development of Spike Island, and for the project that united Kilmainham courthouse and jail into one heritage site. In an ideal world, every museum would have a wealth of engaging tales to choose from, evenly spread across the lifetime of a site, but that's never the case. While researching the history of Spike Island, I came across a selection of fascinating stories about prisoners on the island, but most of them came from the 1850s, and if we had used them all, the exhibition would have placed a disproportionate emphasis on one short period. Museums and historic sites need visitors, and they need to have some big stories to lure them in. This means that if a prison has had a high-profile inmate like John Mitchel, then their story will be told, no matter how fleeting their stay there. Current trends and anniversaries also influence what incidents are highlighted. I was working on Spike Island in the run-up to the centenary of the 1916 Rising, so we included the story of the *Aud*, even though the crew members were held on the island for a matter of hours, because we suspected even a tangential link to the Rising would be popular.

Artefacts are crucial to telling a good story. Without an object to bring them to life, stories are just text on a wall. Museums are shrines to the past, and shrines need relics, no matter how mundane they may seem. In Carlow County Museum there is an artefact so tiny and apparently insignificant that, in life, it never merits more than a glance, and yet it forms the centrepiece of a display. It is the butt of a 'Sweet Afton' cigarette, less than an inch long. In fact, it's burnt so far down that only the 'SWEE' part of the brand name remains. This is all that's left of Kevin Barry's last cigarette. In November 1920, Barry, a young medical student, was executed in Mountjoy Prison for his role in a Volunteer ambush that killed three British soldiers during the War of Independence. Barry was immortalized in the eponymous ballad which praised him as 'Another martyr for old Ireland' and comforted the grieving by assuring them that 'Lads like Barry will free Ireland/ For her sake, they'll live and die.' I find his cigarette butt intriguing. I wonder who picked it up and pocketed it and where it was kept before being donated to the museum. Was it on display in someone's house, or kept in a drawer and brought out only on special occasions? Was it looked at with great reverence, as someone's most precious keepsake? Whatever the answers, it's a powerful example of the way an association with an individual or an event can transform the most mundane object into something very memorable – after all, there are few cigarette butts I would drive 120 kilometres to see.

Tarbert Bridewell in north Kerry was built in 1831 and operated as a small local prison and a petty-sessions courthouse for over a century. The jail was sympathetically restored in the early nineties, though there are occasional imaginative flourishes, such as the iron railings that fan out either side of the entrance posts, which were not there when it was a prison but were added to make the jail feel 'more jail-like'. The Bridewell stands close to the shore of the Shannon Estuary, beside a high stone wall with red-painted doors into which is embedded a restaurant called Spice Village. As if to emphasize the town's cheerful mix of old and new, the prison is encircled on three sides by a modern pebble-dash housing estate. A sideburned Victorian policeman, black uniform patched at the elbow with silver duct tape, stands guard in front of the entrance. We bought our tickets in the café, from a woman who led us to a side door and unlocked it, like a jailer. 'We've just had a new audio system installed,' she explained, 'so give it a minute or two to warm up.' To make the experience more interactive, she picked two cards from a rack on the wall and handed them to us. 'Sure you'll be wanting rid of him,' she said, looking Al up and down before handing him a card revealing that he stood accused of stealing hairnets, ribbons and scissors from a barber shop (given Al's baldness, it seemed an unlikely offence). I was charged with breaking a lamp in St Michael's parish. 'You'll find out what happens at the end,' said the woman ominously, and shut the door behind us.

In the absence of reliable records about the prisoners, the tour follows the story of one 'Thomas Dillon' through the nineteenth-century judicial system in Kerry. Dillon is a composite character; although his story isn't true, it could be. In the first room we found a policeman manhandling Dillon outside a cottage, egged on by a woman in a shawl. After a minute or so, the new audio duly kicks in and we hear the woman, Peig Ahern, accusing Dillon of putting his cow in her field to graze. Though Thomas vigorously denies the charges, we find him, moments later, sitting in one of the cells bemoaning his fate. He is ruined, he groans. His wife and children will starve. In the next room we find the officer recording Dillon's arrest in the prison log book, and the last room presents us with a dramatic courtroom scene in which Dillon – facing vigorous testimony from the relentless Ahern, stout defender of her grass – awaits his inevitable fate. We detour briefly into the exercise yard at the back, where Dillon languishes against a wall in his ruffled white shirt like a Romantic poet in the throes of composing a lament, before returning to the court to learn both Dillon's fate and ours. Thomas Dillon is, somewhat anticlimactically, ordered to pay a fine. Clearly, he will be unable to do so, but the display – concluding, like the magistrates, that this is none of its concern – does not explain what will happen next. As a rookie lamp-breaker, I was ordered to pay a fine of £5 or spend two months in prison. 'And what about him? Did you get rid of him?' asked the woman when we returned to the café, a mischievous glint in her eye. I told her that I did. Al got seven years' transportation. He didn't seem too bothered by the news.

If Dillon had gone to jail, he would have known what to expect. Life in Irish prisons in the nineteenth century was strictly regimented and, though there were some minor differences from prison to prison, the regime was very similar across the country. In 1879 a typical day commenced at 6.30 a.m. with a bell tolling. Exercise began at seven and by eight prisoners had started their work. Work was a key part of prison life. Prison reformers argued that working made confinement easier to bear, it occupied the prisoner and taught them skills they could use when released, and it created revenue for both the prison and the prisoner. When prison goods were sold commercially, one-third of the profits was kept aside for prisoners until their release, when they were presented with a new suit and the money they had earned. Male prisoners were responsible for the maintenance of the building – they painted it, repaired it and looked after the yards. The female prisoners were set to work baking, washing, cleaning cells and sewing. In most jails, prisoners made their own uniforms and shoes. Some prisons sold their produce commercially – in Wicklow Gaol prisoners made fishing nets that were sold in Arklow, a local fishing port, though this was short-lived, for the governor feared the nets would be used by prisoners to fashion their escape over the walls. In Cork City Gaol, clogs, brushes, buckets and mats were made for prison use, while clothing and sheets were sold commercially.

An inspector visiting Nenagh Gaol in the 1850s reported that prisoners completed a wide range of tasks – some of the male prisoners were working as carpenters, cobblers, weavers and tailors, while the women were occupied with needlework, knitting, spinning and carding, though, as in the workhouse, most inmates were doing hard labour, either picking oakum or working in the stone-breaking yard.

At eight thirty breakfast was served, and work then continued until dinner at one. Food was primarily a diet of bread and milk – in Nenagh during the Famine the prisoners were given one pound of bread (the equivalent of a small loaf) for breakfast with a pint of full milk, and for dinner another one pound of bread with a pint of skimmed milk. On Spike Island in the early 1860s meals were characterized by quantity rather than variety. Prisoners received bread and milk three times a day, rice and gruel twice a day, and a portion of meat and a tiny amount of vegetables at dinner – the equivalent of a third of a carrot or a couple of tablespoons of peas. After dinner, any prisoner attending school had lessons for two hours before recommencing work until five forty-five, when they paused to eat supper. Work then resumed until eight and lights were extinguished at eight thirty.

Infractions were heavily punished. An inventory of Wicklow Gaol makes for a sobering read. The jailer had 11 pairs of handcuffs, 21 pairs of bolts with shackles . . . 1 dungeon lock, 2 neck yokes . . . 6 pair of manacles . . . and the branding iron which may have been used to brand John Honige's hand in 1772 after he was found guilty of stealing linen. The treadwheel, invented in 1818 by William Cubit, was used in every county jail until the end of the nineteenth century. It was a never-ending staircase – very similar to the cross-trainers in a gym. Wooden steps were built around a rotating metal frame. As the frame rotated, prisoners were forced to keep stepping up. Occasionally, treadwheels served a purpose – to pump water or grind flour – but in most cases they were simply designed to frustrate and exhaust. There is a replica treadwheel in Wicklow Gaol, though, sadly, visitors can't try it out for themselves.

Flogging was a regular punishment and sometimes formed part of a prisoner's sentence. Michael Reilly was twelve years old in 1833, when he was jailed for stealing three ducks and two hens. He was sentenced to three weeks in Kilmainham Gaol and to be whipped once a week – twenty lashes each time. Most prisons also had the crank, which was a large drum or tombola with a crank handle. There is a replica in Crumlin Road Gaol, which looks like something that might be used to draw out raffle tickets, though it's much less benign. The handle was attached to cogs inside the drum and, as part of the punishment, prisoners turned it a set number of times. Prison officers could tighten the cogs to make it harder to turn. The 'shot-drill' required a prisoner to bend down while keeping their legs straight and pick up a heavy steel ball (like a cannonball) and raise it to chest height. There was no purpose to any of these punishments; they were designed simply to torture. If the architecture of prisons encouraged prisoners to raise their eyes to heaven, their bodies remained firmly trapped in hell, performing endless Sisyphean tasks designed to dehumanize rather than reform.

Almost all prison sites have stocks or pillories (stocks trapped seated prisoners' ankles between wooden planks; pillories secured their neck and wrists). Like the treadwheel and cranks, they're replicas, and they would not have been used at the prisons – generally, they were set up in town squares, where the public could hurl rotten fruit, vegetables and anything else they could get their hands on at the unfortunates, who had often been found guilty of public drunkenness. The replicas you find at former prisons are there not to bolster the immersive authenticity of the site but to provide an Instagram moment for visitors posing in them (fruit and vegetables are generally not provided).

There were four ways to get out of prison. Transportation (that is, swapping one form of incarceration for another), release at the end of a sentence, escape and execution. Transportation of men, women and children took place in Ireland for over two hundred years from the 1620s to the 1860s. Although some were transported for political reasons, by far the most were convicted of crimes which ranged from the petty – begging and stealing potatoes – to murder. The most documented period is from the 1790s to the 1850s, when the vast majority of those sent from Ireland went to New South Wales or Van Diemen's Land (known as Tasmania from 1856).

Those awaiting transportation were funnelled through the prison system until they arrived at convict depots such as Spike Island, where they were held until they boarded ships bound for Australia. The story of transportation is most vividly told in the cells and corridors of the prison museums, including Kilmainham Gaol, Downpatrick Gaol, Elizabeth

Fort and particularly Spike Island, which from 1847 was the main transportation depot.

The state embraced transportation for many reasons. It was sometimes used as a substitute for death sentences, but more often those dispatched to a foreign land were poor people found guilty of very minor offences and the severe sentence was intended to deter potential criminals. About five thousand, or 12.5 per cent, of all transported prisoners were political, and the largest number of these were prisoners associated with the 1798 Rebellion. But above all transportation was a crucial part of the colonial project, providing the colonies with free labour.

There are detailed records for the forty thousand men, women and children transported to Australia between 1791 and 1853. They make for shocking reading: Jane Armstrong, aged ten, transported for stealing two silver spoons; Mary Acheson, aged sixty-two, transported for stealing geese; fourteen-year-old James Loughran, transported for begging; Michael Fitzpatrick, aged twenty-three, transported for stealing potatoes; fifteen-year-old Thomas O'Neill, transported for stealing three bridles; Murphy Callaghan, aged twenty-seven, transported for stealing books; James Cleary, twenty-eight, found guilty of bigamy and sentenced to seven years' transportation in 1846. Only about 5 per cent of those transported returned to Ireland after they had competed their sentence (which was for seven or fourteen years, or banishment for life). This was partly because the fare back was too expensive, but also because, like the indentured servants of old, many discovered there were greater opportunities available to them in their new home.

Hidden inside the former Garda station on the old parade ground of Elizabeth Fort in Cork is a small exhibition, *Walls, Women, Water* (it seems someone was desperate for an alliterative title). Disappointingly, there are no artefacts, but a series of panel texts briefly outline the history of the fort before focussing on the 1820s and '30s, when the fort was a convict depot for women who had been sentenced to transportation. Women and girls made up almost a quarter of the forty thousand convicts dispatched to Australia. Of all the women transported, the Irish were disproportionately represented, with Deborah Oxley estimating that 56 per cent of female convicts were Irish and only 34 per cent English (the rest came from Scotland and Wales). Many were first-time offenders, often women who had moved from the countryside to the city in search of elusive work. While in Elizabeth Fort, the women made clothing packages for their long voyage – every woman left the fort with a woollen jacket, two shifts (loose-fitting dresses), two linen caps, a straw bonnet, a check apron, two handkerchiefs, a pair of stockings, a pair of shoes and a linen bag. They also made prison uniforms and suits which were distributed to male prisoners in other jails. In 1837 the fort was supplying a thousand suits a year to convicts.

Most of the women transported had been found guilty of theft, women like Mary Cassidy, aged nineteen, who was sentenced to seven years for stealing a handkerchief, and Mary Kerrigan, for theft of a tablecloth. In 1833, Mary Russell was charged with stealing a piece of cloth and also sentenced to seven years' transportation. She was taken to Cork County Gaol (now demolished – though the boundary wall and gateway to the prison form part of the UCC campus), and John, her six-year-old son, was left to fend for himself. Under this Dickensian system there was no official concern about what would happen to the boy; he was simply abandoned to walk the streets. How he survived, we don't know, but three months later he was arrested, charged with vagrancy and reunited with his mother in the prison. The story does not end happily there, or anywhere. Both mother and child were moved to Elizabeth Fort and in January 1835 they boarded the *Neva* and sailed for Australia. On 13 May the ship hit a reef near Tasmania, and almost everyone on board, including Mary and John, drowned.

The male convicts who made it to Australia alive generally served their time as farm labourers or were put to work constructing the roads, bridges and buildings of the new colony. Deciding what to do with female convicts was more problematic. Some were sent to work as domestic servants, but most were sent to Parramatta Female Factory, which was a mix of factory, prison and workhouse about twenty-five kilometres west of Sydney. There, women were divided into three classes, according to the severity of their crime. The first-class prisoners were paid a low wage and worked as weavers, seamstresses and hat makers. They could leave if they found a husband or a job. Those in the second class were also paid and worked in the clothing industry, but they were confined to the factory, while third-class prisoners were unpaid, had their heads shaved and broke stones in the yards or picked oakum.

In the 1850s, transportation was slowly wound up, partly because thousands of regular emigrants were moving to Australia and the authorities no longer desired convict labour to build infrastructure and work on the land, and partly because transportation was an expensive way of dealing with minor criminals. The last convict ship to sail to Australia was the *Hougoumont*, which arrived in Freemantle in Western Australia in January 1868 with 280 convicts on board, including 62 Fenians. Eight years later, six Fenian prisoners escaped from Freemantle in dramatic fashion, boarding a rescue ship, the *Catalpa*, and sailing to New York, where they were met with a hero's welcome.

Like the Fenians, most prisoners were keen to get out before their sentence was completed. Escapes from prison are the stuff of Hollywood films, though – sadly for this book – the reality is that most attempts lacked the great drama of a blockbuster. Many of them were organized with the assistance of the warders. Indeed, in August 1801, nine prisoners escaped from Kilmainham Gaol by the ingenious method of walking out the front door. In October 1921, while Spike Island was being used as an internment camp during the War of Independence, six IRA men did escape in dramatic fashion as one of them, Henry O'Mahony, later recalled:

> We tunnelled through a wall surrounding the prison . . . We then scaled the outside wall by means of a timber ladder – made from the joists of the flooring of the prison . . . We made our way to the coast and eventually to the pier, where we [found a rowing boat] . . . into which we tumbled, and with the aid of a storm, succeeded in reaching Cobh and safety.

In Belfast, Crumlin Road escapees favoured the over-the-wall approach, with prisoners escaping in 1943 by standing three men high on each other's shoulders like a circus act. In 1960, Donal Donnelly escaped over that same perimeter wall using knotted bedsheets and electric flex to make a rope seventy feet long, while in 1971 nine members of the Provisional IRA used rope ladders to escape. Two days later, two of the men were recaptured dressed as priests and accompanied by two (genuine) monks who were later charged with abetting them.

The fourth and final way out of prison was the least popular. Between 1796 and 1924, more than 160 executions took place at Kilmainham Gaol, far more than at any of the other sites now open to the public. A number of those executed were political prisoners, including five members of the 'Invincibles' (a splinter group of the IRB) executed in 1883 for their role in the murder of Chief Secretary Lord Frederick Cavendish and Under-Secretary Thomas Burke in Phoenix Park. But, as at every prison, most of those killed were executed for non-political crimes ranging from theft to forgery to murder.

In 1868, following a prolonged campaign by prison reformers and those opposed to the death penalty, the Capital Punishment Amendment Act banned public executions, though they continued to be held within the confines of prison walls. Prior to 1868, executions were carried out in public, often from gallows erected at the front of the prison, but sometimes freestanding gallows were erected in town squares. Above the entrance to both Kilmainham and Down County jails, bricks embedded in the limestone indicate where the execution platform and gallows once stood. The executions provided a macabre form of public entertainment as thousands flocked to these gruesome events.

At Nenagh Gaol the spot where the gallows once stood is now obscured by a life-size plaster statue of the Virgin Mary. She stands at the centre of a high arch which forms part of the prison gatehouse. I passed beneath her and headed down a long avenue flanked by twenty-five-foot-high walls to Nenagh Heritage Centre, which is housed in a large octagonal building that was once the Governor's House (when the prison closed in 1887 it became first a convent and later a school, which explains the statue of Mary). Inside the heritage centre, there's a scale model of the jail as it was when it opened in 1842, showing the seven prison blocks that used to radiate from the Governor's House – only one of these remains today. I learned about William and Daniel Cormack, who, in May 1858, were the last men executed in Nenagh Gaol. The brothers had been found guilty of murdering land agent John Ellis, but there was widespread outrage at their conviction, as it was believed that the evidence against them had been secured by bribery and threats and several witnesses later admitted that they had given false evidence to prevent themselves being accused. Before the executions, over two thousand people signed a petition seeking leniency, but to no avail.

Between 1842 and 1858, seventeen men were executed in Nenagh, all for conspiracy to murder, attempted murder or murder. The Cormack brothers were held in separate cells in the gatehouse, each with access to its own tiny exercise yard – sixteen foot square – surrounded by twenty-three-foot-high walls on every side. To walk a mile, a prisoner had to lap the yard seventy-five times. I stood in the yard and looked up at what Oscar Wilde called 'that little tent of blue/ Which prisoners call the sky'. As a twelve-year-old, I'd recited several verses of Wilde's 'The Ballad of Reading Gaol' in an old school hall. I don't recall why this happened, but I have a very clear image of standing on a small wooden stage facing a room of adults sitting on plastic seats listening to me intoning, 'Yet each man kills the thing he loves . . .' Clearly, I've been collecting material for this book for a very long time.

On 11 May 1858, the Cormack brothers were taken from their cells to a narrow corridor spanning the arch. To their left was the execution room, to their right the scaffold. In the execution room the brothers had their hands tied and a white cap placed over their heads, before they were led out through the arch (past where the statue of Mary now stands) to the gallows. The *Tipperary Advocate* reported that William Cormack 'kissed the crucifix he held in his hand, while the miserable hangman was adjusting the pinions'. Several thousand had gathered to bear witness to the executions and a ballad written at the time lamented:

> The day of their execution, as they stood on the drop,
> The thunder came so dreadful that it did the people shock;
> It seems the Lord was angry, in which He showed His power,
> As they were dying innocent upon that dismal hour.

As always, a ballad may have caught the emotion, if not the facts, for the *Tipperary Examiner* reported that 'the day was beautifully fine, and the sun shone, I trust, on the innocence of those poor men'. Several weeks after the Cormacks were executed a crowd of about fourteen thousand assembled in Nenagh to protest against the injustice, and the campaign continued until finally, in 1910, the bodies of William and Daniel Cormack were exhumed and buried in a mausoleum at their home cemetery in Loughmore, thirty-five kilometres away. An estimated ten thousand people lined the streets of Nenagh to watch the cortège pass, accompanied by twenty bands and several hundred jaunting cars.

Of all the places I've visited, execution chambers are the most disturbing, and it is the one in Mountjoy Prison that sticks most clearly in my mind. I had gone to visit the museum, which – since Mountjoy is still a working prison – sat slightly outside the grounds, housed in an old cash-and-carry warehouse (it recently moved to a new site). The museum exists largely because of the determination of two prison officers: Jim Petherbridge and Seán Reynolds. Seán, a genial, quietly spoken man in his seventies, has devoted much of his retirement to curating and acting as the custodian of the eclectic but fascinating collection which includes leg irons, an executioner's dummy, a rope made of bedsheets (Donal Donnelly was not the only prisoner to plan such an escape) and a dictionary hollowed out to hide a mobile phone. As I wandered around the exhibits Seán casually asked, 'Would you like to see the hang house?' Without giving it any thought, I eagerly agreed. It was only then that I realized that the hang house was inside the prison.

Historic sites bring the past closer. It's easier to imagine knights clanking about in their armour when you're standing in a castle, or prisoners breaking stone when you're standing in a prison yard. And yet there is often something filmic about it, as if these people weren't quite real. Mountjoy – a working prison – isn't like that. Going through security, taking off my jacket and belt, emptying my pockets and leaving behind both phone and bags, makes it all feel much more immediate and real. I followed Seán into the main part of the prison and with a jolt I realized that the radial layout was almost exactly the same as that of Crumlin Road Gaol in Belfast, but instead of tourists, mannequins and guides, there were prisoners and prison officers. Mountjoy Prison is 170 years old, and while the building may have represented the best of reform architecture in its time, today it feels anachronistic, and there's something unsettling about the fact that it's still being used as a place of incarceration. I walked quickly behind Seán, trying not to catch anyone's eye, and followed him into the hang house.

The hang house is a brick, whitewashed room with no windows. It was here that forty-five men and one woman were executed between 1901 and 1954, including Kevin Barry and nine other volunteers during the War of Independence. Capital punishment wasn't formally abolished in Ireland until 1990 (it was abolished in Northern Ireland in 1973). Glass panels in the pitched roof let in natural light and offered the condemned a final glimpse of sky. At the centre of the room is the trapdoor (now surrounded by a wooden railing) and above it is a joist from which hangs an iron chain. There is something very unsettling about being in a room that was built with the express purpose of killing people. It's one thing to be somewhere where people died, but places built to cause intentional death are, to my mind, abhorrent.

There's an iron lever at one side of the trapdoor.

'Do you want to pull it?' Seán asked.

This was my second execution chamber in as many weeks. The first, in Crumlin Road Gaol, was also distressing, in part because of the behaviour of some members of our tour group. Rather bizarrely, the execution chamber is accessed through the en suite bathroom of the condemned man's cell, and we shuffled into the small room and fanned out around the edges, keeping well away from the trapdoor in the centre. As with Mountjoy, above the trapdoor was the gallows and from it hung a noose. I ended up surrounded by a group of men in their thirties, a stag party that cracked jokes about suicide, murder and lynchings. Through their giggles and snorts I tried to listen to the guide as he talked about the seventeen men who were executed in the prison. Their names, projected on to the wall behind the gallows and noose, scrolled past as the stag party tried to line up for selfies before being admonished by the guide. My most charitable take on their behaviour is that they were uncomfortable and covering it with crass bravado. Now, standing in the empty hang house in Mountjoy, I shivered and declined Seán's offer. Seán pulled the lever himself and the trapdoor swung open with a jolt, revealing a room below with a set of mobile wooden steps used to release the dead prisoner from the noose.

Back in the museum, Seán pointed out the executioner's box,

full of all the paraphernalia of death – a hood, leather restraints and a noose. On the wall was a plaster death mask of Albert Pierrepoint, who had carried out the last execution in Mountjoy Prison, in 1954. Generally, there was no resident executioner in Ireland, and executioners like William Marwood and several generations of the Pierrepoint family were brought over from England to carry out executions when required. When Marwood died in 1883, over eight hundred people applied to replace him. Most had no experience. One applicant wrote: 'I am willing to be elected as a candidate for this situation as I am a butcher by trade . . . I am experienced in tying knots with rapiers.' He didn't get the job. Instead, Marwood was replaced by Bartholomew Binns, but he was sacked a year later following several botched executions and the suspicion that he had been drunk when carrying them out.

Not all executioners were imported. For a few years around the turn of the nineteenth century there was one who was not only resident in Ireland, but a prisoner herself. Elizabeth Sugrue ('Lady Betty') became the executioner at Roscommon Gaol in bizarre (and tragic) circumstances. The story of Lady Betty has a number of versions, but a combination of an account by William Wilde (Oscar's father) and the stories from the Irish Folklore Collection yields an interesting, if possibly embellished, tale. Sugrue was from Kerry but settled in the ominously named Gallowstown just outside Roscommon Town. She was well known for her violent temper, and her son left as soon as he was old enough to enlist in the army. Some years later, he returned and took lodgings in his mother's house without revealing who he was. Though his mother somehow failed to recognize him, she could see that he was a man of means and, as he slept, she murdered him and stole his money, only later realizing she had killed her own son. She was arrested, tried, sentenced to death and scheduled to be executed alongside several others. The date of execution duly arrived, but the executioner did not. In what was either a piece of brilliant opportunism or a burst of enthusiasm for public service, Sugrue volunteered to carry out the executions herself, in return for her life. Her request was granted, and for several years she not only acted as executioner but also carried out public floggings in the market square. In her rooms in the jail she used a burnt stick to draw portraits on the walls of all the prisoners she'd executed. But executioners are not traditionally popular, and in 1807 she was killed when prisoners in the stone-breaking yard – perhaps hoping to usurp her place – used their stones to beat her to death.

Mountjoy Prison is one of the few Victorian jails still in use today. In 1796 there were fifty-one jails in Ireland, but by the late 1870s that figure had risen to a hundred and fifty – mostly small bridewells which held prisoners awaiting trial, but also county jails, convict depots and debtors' prisons. By 1878, these jails held forty-five thousand prisoners, one-third of them women. Today, there are fifteen prisons on the island, housing just under five and a half thousand people (just over two hundred of whom are women). There's no need for most of the nineteenth-century prisons, and in many cases the buildings have been at least partly demolished. But as with workhouses and asylums, some prisons have acquired a new life: the former prison in Armagh is ear-marked to be renovated into a luxury hotel, while in Sligo St Columba's psychiatric hospital is now a hotel, though all that really remains of the former asylum is the imposing grey lime-stone façade. The hospital closed in 1992 and became derelict, with several fires destroying much of the interior, and when I stayed there I didn't notice anything that acknowledged the building's difficult past. The Ambassador Hotel on Military Hill in Cork was formerly the Home for Protestant Incurables and there's an institutional feel to the wide corridors and balconies that were intended to allow patients to get some fresh air. In Drogheda, the perimeter wall of the old jail encloses Eddie's Hardware, while in Dundalk the prison has been repurposed as the Oriel Centre, a performance and exhibition space.

The imposing exteriors of Roscommon and Carlow jails now form the entrances to shopping centres (because nothing sets the tone for a successful retail experience like an imposing 'abandon hope'-style façade). If you glance up as you approach the entrance, you can just about see where the gallows used to be – in Carlow's case, the mark in the keystone is obscured not by a life-size statue of Mary but by the two blue plastic 'P's in the 'Carlow Shopping Centre' sign that curves around the arch. Just below the large window that looks out on to Kennedy Avenue you can see where concrete has filled in the gap made by the removal of the execution platform. Inside, the prison yard is enclosed by a glass roof and the Governor's House now serves tea and scones, but there are still traces of the building's previous incarnation and, as I looked around for the entrance to Penney's, I could see cell windows, complete with iron bars.

While some buildings are completely transformed and repurposed, others commemorate their original use. Many of these are private enterprises under pressure to turn a profit and seek out commercial opportunities. Indeed, these former prisons are eager to host everything from gala banquets to gigs. Crumlin Road Gaol hosts a New Year's Eve Governor's Ball. Lifford Courthouse uses its cells as escape rooms for stag, hen and birthday parties, while Belfast and Wicklow jails run paranormal tours. Cork City and Crumlin Road jails promote themselves as ideal wedding venues, putting a whole new spin on the phrase 'ball and chain'.

The exception is Kilmainham Gaol, which occupies a unique, sanctified space in the Irish imagination and identity. There are no ghost tours or wedding parties there. As a prison which held many Irish nationalist heroes and, most significantly, as the site of execution of the leaders of the 1916 Rising, it cannot be seen to trivialize (or commercialize) the hallowed ground on which their deaths took place. Yet Kilmainham's conversion to a – if not the – key site associated with the Irish Revolution was far from immediate, and for decades after the last prisoner left in 1924 the site was abandoned and left to rot. It seemed the state didn't know what to do with such a politically sensitive location. In the late thirties, it was suggested that the buildings be demolished and a children's playground built. Later there were proposals for a military museum or a memorial garden. In the end, locals and former prisoners took matters into their own hands. In 1958 the Kilmainham Jail Restoration Committee was formed and sought public donations for the restoration. In 1960 the committee was given the keys to the jail and work began. By then the site was virtually derelict, colonized by pigeons and rats, with trees growing thirty feet high inside the prison and brambles and bushes carpeting the yards. Hundreds of volunteers – many of them veterans of the revolutionary period who had been imprisoned in the jail – gave their time, energy and expertise to help restore it. On Easter Sunday 1966, as part of the fiftieth-anniversary celebrations of the 1916 Rising, President Éamon de Valera (who had been one of the last prisoners to leave the jail in 1924) returned to lay a wreath in the yard where the 1916 leaders were executed.

A nuanced reflection on the history of incarceration and of all those held in Kilmainham Gaol is now told in the museum and on tours. But it wasn't always the case. Those who took on the mammoth task of restoring the jail did so with a particular goal in mind: to 'perpetuate the memory of the illustrious dead'. The volunteers restoring the building were told to 'enter the jail as you would a shrine'. In 1965, Lorcan Leonard, one of the key drivers behind the restoration, described the jail as the 'Calvary of republicanism in Ireland', and from the outset the committee planned to use it to tell the story of Irish republicanism.

In Kilmainham Gaol, as at many other museums and historic sites, the story of 1916 flows into the story of the War of Independence, and there are two objects on display in the museum which illustrate both the callousness and the humanity of conflict. One is a small metal HMV box, complete with the iconic image of a dog with its head cocked listening to music from a gramophone, which held several bullet fragments prised from a kitchen dresser. The bullets had killed George Ennis as he stood in his kitchen. They had been fired by one of the British soldiers who, on 29 April 1916, had rampaged through North King Street in the inner city, brutally murdering fifteen men and boys, none of whom were involved in the Rising. Despite a secret military inquiry about the incident, no action was taken against the troops or their commanders. In another exhibition case there is a box of Orchid Chocolates, still in its cellophane wrapper. The chocolates were a gift to Thomas Whelan from Captain Lester Collins, a Black and Tan who had befriended Whelan while he was in Mountjoy Prison awaiting execution for his alleged involvement in the death of several British intelligence officers in November 1920. Whelan sent them to a friend, promising that if he was reprieved, they could eat them together; if he was not, she should eat them all. Whelan was executed on 14 March 1921. The chocolates were never opened.

Until relatively recently, tours of Kilmainham Gaol offered only a very cursory glimpse into the Civil War. When I started visiting the prison, the tour would conclude with stories about the War of Independence but often make no mention at all of the Civil War, even as we passed a small plaque on a wall of one of the exercise yards that displays the names of four anti-treaty men who were executed there in November 1922 – they were the final executions to take place at the jail. Now there is a much greater emphasis on the Civil War, in part because it offers an opportunity to tell the stories of some of the almost five hundred women who were imprisoned there in 1923. Graffiti on cell walls bears witness to their time there, an experience Dorothy Macardle, one of the republican prisoners, recalled as one where she had 'the sense of accumulated tragedy, endless sacrifice, the never ceasing persecution of those who stood for Ireland's freedom ... The very stones of our exercise yard speak to us of the blood that stained them in that awful week of May 1916'. Autograph books, a series of satirical cartoons by Constance Markievicz and a chair leg used by women prisoners as a rounders bat are on display in the museum. Displaying these objects and telling stories associated with women prisoners provides necessary reminders that women as well as men played significant roles in key events – something that was often overlooked until relatively recently.

The jail's archival collection is heavily skewed towards one version of Irish republicanism. Many of those who donated material to the jail, particularly before it was handed back to state control in 1986, came from families who had opposed the Anglo-Irish Treaty and fought against it in the Civil War. These families often wanted little to do with the Irish Free State, later the Republic, and while they wanted their family members' stories to be told, they did not wish to leave papers and artefacts to the National Museum – a state institution. Today, the jail's archive still reflects that tension, with 54 per cent of the collection relating to the period between 1916 and 1924, and over 80 per cent of the archive relating to Irish nationalist and republican history. This is understandable, for political prisoners are far more likely to have a collection of papers to donate, and the families of non-political prisoners are unlikely to want to perpetuate the memory of their ancestor by giving material that can be used by researchers and in the museum displays. In fact, there is only one artefact in the entire Kilmainham collection that is directly connected to an ordinary prisoner. It's a service medal presented to Henry Rance for his efforts in repulsing the attempted Fenian invasion of Canada in 1867. Later, Rance settled in Ireland and was imprisoned twice in Kilmainham Gaol – the first time in 1892 for threatening his wife, the second time a year later for attempting suicide (which was a crime in Ireland until 1993).

When I was about ten years old, I visited the prison for the first time. I was clearly a melancholy child, what with my recitation of poems about incarceration and death and my desire to wander around old prisons. I don't remember much about that tour except for the story of Anne Devlin, who, in the aftermath of the 1803 Rebellion, was arrested, imprisoned and tortured in an unsuccessful attempt to get her to reveal the hiding place of Robert Emmet, the leader of the rebellion. I vividly recall standing in the prison yard while our guide pointed out a small, barred window just peeking above ground. 'That's where she was held,' he announced. I was horrified and fascinated. Anne Devlin wasn't mentioned on all the tours I took in subsequent years, but each time I stood in the yard I looked at that cell window and imagined the terror she must have felt. Nearly thirty years later, while I was working on a heritage project at the jail,

I learned that there are no cell allocation records for that period and nobody knows where she was held. She might have been in that cell, but she probably wasn't. It makes no real difference to the story of Anne Devlin's horrific time in the prison, but I still felt a jolt of disappointment that for so many years I'd been paying silent homage to her at the wrong place.

When a building has undergone many changes of use over time, or has fallen into disrepair, it's impossible to re-create it as it was 'in the past'. Which past? One site that embraces this impossibility with aplomb is 14 Henrietta Street in Dublin. The street was once one of the most exclusive and fashionable in the city, home to lords, ladies and bishops. A century later it was a slum, and today visitors to the house are regaled both with tales of Georgian splendour and stories from the later history of the house, which, by 1911, was home to more than a hundred people living in nineteen flats sharing one tap and one toilet in the yard. Delicate Georgian stucco and fine panelling sit cheek by jowl with walls painted in the characteristic tenement paint, Reckitt's Blue and Red Raddle. It's a beautifully evocative tour which uses the building itself as its most important artefact.

At Kilmainham, much of the site remains as it was when prisoners were held there (or at least those held there from the 1860s onwards). But, like all historic sites, there are gaps in our knowledge that make a completely authentic restoration, if such a thing exists, impossible, even without the modern interventions required to provide heat, light, access routes and assembly spaces. And there are places where Kilmainham Gaol is more reproduction than restoration. All tour groups pass through a narrow doorway to the so-called '1916 Corridor'. Those who look above the doorway will see a piece of graffiti illuminated by one bare bulb. It reads, 'Beware of the Risen People. They that have harried and held. Ye that have bullied and bribed.' These words of Patrick Pearse were probably painted there by prisoners during the War of Independence, but were later removed because guides believed they dated from the sixties, when the restoration work was carried out. When former prisoners recalled seeing them while imprisoned, the words were repainted. The authentic was believed to be fake, and the fake is now presented as authentic.

There's a similar story in the East Wing. On every tour I've taken in Kilmainham, visitors are directed to the cell occupied by the artist Grace Gifford during the Civil War. Here we're shown the painting of the Madonna and child that Gifford painted on the wall of her cell in 1923. Except that it's not the original painting. In the sixties, the artist Thomas Ryan (later president of the Royal Hibernian Academy) traced Gifford's faded and damaged original before the wall was replastered. Ryan then painted a copy of the painting on the wall, adding his own flourishes, using oil rather than the original watercolour and (in a tell-tale move) even signing the painting with his own name. In a letter written in 2005, Ryan explained that his painting 'is not the tentative pencil drawing of the prisoner. No prison governor, no matter how lax the regime, would allow an inmate to come into a cell with a box of oil paints, oil, turpentine and brushes. She used what she had – a pencil and a little box of watercolours.' Despite this admission, Ryan cheerfully concluded that the painting, his painting, even in its second-hand mode, is 'charming'. Yet these stories don't diminish the experience, rather they highlight how historic sites change over time, how most stories are multilayered (literally, in Gifford's case).

These examples were to the forefront of my mind when in May 2015 I stepped on to a small boat and made the fifteen-minute journey to Spike Island. It was my first visit to the island, and I couldn't wait to see it. A fortified island off the coast of Cobh might not seem the most obvious place to develop a tourist attraction, but it isn't the first island prison to open to the public, for there are some extremely successful precedents, including Robben Island, off the coast of Cape Town, and Alcatraz, off San Francisco. Cork County Council, who run Spike Island, envisaged it becoming 'Ireland's Alcatraz'. My job was to tell four key stories – the island's past as military fort, as prison, as home and its relationship with Cork Harbour.

From the pier we walked past an adventure centre and some derelict cottages up a steep hill to the fort. Construction of the fort began in 1804, during the Napoleonic Wars, when Britain feared that the French might invade Ireland, but it took more than sixty years to complete, by which time a design that had once been the pinnacle of defensive military architecture was outdated and in fact the fort had been a prison for nearly twenty years. We passed over the dry moat and through the imposing arch into the huge yard encircled by high walls and bastions. The buildings within the fort have, over the course of two centuries, been home to many, including British and Irish soldiers, rebels, convicts awaiting transportation and archaeology students who spent their summers excavating the prison cemetery.

The first prisoners arrived on Spike Island in 1847 and were housed in buildings that had been intended for use as military barracks, and it was some of these buildings that had been earmarked for development as museum and exhibition spaces. The 'separate system' was not used on Spike, and up to thirty men were crowded into dormitories as small as forty feet by eight (about the size of a badminton court). Their beds were soon replaced by hammocks so that every dormitory could also fit a table that prisoners could eat and work at. For six years, most prisoners spent short periods on the island before transportation but, following the Penal Servitude Act of 1853, the seven-year sentence (the most common transportation sentence) was abolished. Instead, offenders were sentenced to 'penal servitude' in Irish prisons, and the prison population swelled. Spike was soon overcrowded and became the largest convict prison in Britain and Ireland, housing over 2,300 prisoners. Charles Gibson, the Presbyterian chaplain, complained that 'the prisoners are separated from each other by thin boarded and wired partitions, like a menagerie of wild animals, that snarl and fight in defiance of their keepers'.

How do you begin to restore a ruin? Where do you start? I got a sense of what it must have been like for Lorcan Leonard and the other Kilmainham volunteers – the scale of the challenge they faced – when I arrived at the heart of the site. The two buildings where most of the exhibition was to be displayed were the shell store and the prison block. Both had been important buildings in the history of Spike. The shell store has had multiple incarnations – built to house explosives, it was later used as a hospital, dormitories for a hundred prisoners, a store and, finally, an engine room for the island's electricity supply. We had to don hi-vis jackets, hard hats and steel-capped boots and arm ourselves with torches before entering the store, as most of the windows had been bricked up. When we walked in, the beams of our torches lit up rooms full of detritus, with rotting wooden partitions listing to one side and smashed panels of glass everywhere. The space was cold and damp, so damp that white calcium carbonate stalactites and stalagmites were growing on every surface. Clearly, some degree of imagination was going to be required both to see what the building had been and what it could become.

The prison block had a more straightforward story. Despite the thousands of prisoners held on Spike Island between 1847 and 2004, it is the only purpose-built prison block in the fort and was used to house prisoners and later for storage. It's an L-shaped block constructed from limestone containing twenty-eight cells – most had a small window set about ten feet above ground. Some had no window at all and were used as punishment cells where prisoners were deprived of both sound and light for days and sometimes weeks. It was constructed following the murder of warder William Reddy in 1856 when he intervened in a row between prisoners in one of the large prison dormitories. Reddy was beaten to death with iron bars wielded by two convicts, Edmond Power and Patrick Norris, who were serving time for arson and the theft of a cow. Opened in 1860, its first occupants were 'penal class' prisoners, considered the most dangerous on the island, heavily chained and clothed in black from head to toe, with a veil hiding all but their eyes.

Prisoners covered the walls with graffiti and etched their names into the plaster. John L. O'Sullivan, who was held there during the War of Independence, recalled drawing comfort from the names of the Fenian prisoners carved into the walls. Though some cells were knocked together to make bigger rooms and other alterations occurred, the building changed little for 150 years. But in an act of senseless destruction sometime in the 2000s, after the prison had closed all the plaster was removed, exposing the brick beneath, which meant that (unlike in Kilmainham) little evidence remained of those who were held captive there.

Today, the building is used to explore the history of the fort as a prison, from the first convicts arriving in 1847 to the last prisoners leaving in 2004. Several cells have been furnished as they would have been in different periods and there are models of prisoners in some. A soundtrack plays along one corridor, replicating the harrowing noises that might have been heard there, and one of the punishment cells has been left untouched so visitors can enter and close the cell door behind them, experiencing partially, and very briefly, the sort of dank terror a prisoner would have felt in the 1860s.

And, as in Kilmainham, old stories have been revised and retold. In 1938, Éamon de Valera visited Spike Island and was shown John Mitchel's prison cell, furnished as it would have been when Mitchel was there in 1848. The cell the Taoiseach saw was in the prison block which was later renamed the Mitchel Block, but Mitchel was never there, for it was built more than a decade after his brief incarceration on the island. It's likely that he was held in a former barracks building that had been transformed into prisoner accommodation and in 1915 was destroyed in a fire, though the shell still stands. When I was writing the content for Spike Island, I decided to tell the story of de Valera's visit to 'Mitchel's cell', explaining how that story has changed over time as more research has been carried out and our knowledge of Spike Island's past continues to expand.

But there are always more stories that could be told. Buildings tell tales if we know how to read them, but too often the panels describing architectural features are written by experts who use a barrage of architectural terms unfamiliar to most visitors – mullions, lancet windows, triple sedilias, rood screens, quoins, corbels and garderobes – without ever explaining what they are. But when the context is made clear and the language is accessible, these details really bring a site to life. There is one detail I would have liked to incorporate into an exhibition panel in the prison block, but I only learned about it later. Soon after the official opening of the exhibition I took a trip out to Spike with Jessie. Long before I began working on the island, she had written a report on the historic buildings within the fort, and as we entered the prison building she stopped and pointed out that every limestone block was scored with a unique set of decorative marks, the sort you might see on an archway, around the entrance or windows of a fine house or public building, but not on a prison block. I'd never noticed these marks before because, though I'd seen the building many times, I'd never really looked at it. Prisoners on Spike Island built much of the naval shipyard on neighbouring Haulbowline, and also constructed the prison block. Many were trained as stonemasons, and every chisel mark on the limestone blocks of the prison building was made by a prisoner. So little evidence of the prisoners who were held on Spike in the nineteenth century exists that there's something poignant about seeing these marks. The graffiti in the cells may have been destroyed, but fragments of the lives of those who were held on the island remain and now, every time I visit that prison, I think about those prisoners who built the walls that enclosed them.

Back in the yard of Glenties Courthouse, a steady rain continued as I considered our inadvertent incarceration. I was sure that the situation would be easily remedied. All I had to do was call the front desk. I found the museum website on my phone and called the number, and it was only when it rang out unanswered that I became a little concerned. I tried again, and then a third time, with no reply. The rain began to soak through our coats. We tried banging on the fire-escape door, but no one came so we hurried over to the wooden gate between the museum and the courthouse. The gate was eight feet high and topped with barbed wire. We put our faces to the slats and peered into the street, watching for passers-by to alert, but there were no passers-by. By now it was well after four o'clock. Surely the museum staff had not forgotten us and gone home? I tried the number again, to no avail, while the rain turned increasingly immersive.

Until now, Al had been a reluctant and sometimes recalcitrant companion on my tour of the various miseries inflicted by his people on mine, but in a crisis he has been known to spring into action, and now, with the aid of some disused pallets, a wobbly chair and a side order of desperation, he scrambled up the soaked and slippery gate, hoisted himself gingerly over the barbed wire and dropped down on to the footpath unscathed before disappearing into the museum to negotiate my release. Inside, he found two volunteers deep in conversation. He explained patiently what had happened, expecting – not unreasonably – a degree of mortification and fuss to ensue, but they were unmoved. Why hadn't we called the front desk? they wanted to know. Al pointed out that we had called the number on the website several times. 'Ach sure, that's the wrong number, we've no idea where that rings,' said one of them. Since this appeared to be the end of the conversation, Al took himself back downstairs and let me in through the fire exit. I trudged back in, looking like a drowned banshee, with murder in my eyes. 'Oh, there you are,' said one of the volunteers. 'Would you just sign the register? And would you mind not mentioning what happened?' 'Of course,' I replied. 'I won't say a word.'

Beyond the Land

One place that has everything you could wish for in a museum of Irish misery – poverty, famine, eviction, suffering children, injury and emigration – is the Michael Davitt Museum in Straide, County Mayo. The museum is housed in the modest whitewashed chapel where Davitt was christened, a few yards from the field where his family home once stood (the exact location depends on which farmer you talk to) and from which he was evicted as a small child, one of several traumatic early experiences that shaped his future career. Al and I pulled into the car park, slightly puzzled by the sight of a Chinese flag flapping proudly from a pole. We were welcomed by Tracy, a slight young woman with square glasses and dark hair tied in a ponytail, and Cathy, a silver-haired woman from Lancashire whose 'party trick', she told us laughingly, was that she sounded like Davitt. Tracey ushered us over to the AV near the altar and put on a short introductory video about Davitt's extraordinary life, in the middle of which she reappeared with two cups of tea and a selection of biscuits, immediately bumping the museum into my all-time top ten. After their eviction, Davitt's family set sail for England, which is like being mauled by an escaped lion and then moving in with the pride. They settled in Haslingden in Lancashire, where young Davitt was put to work in a Dickensian cotton mill. There he suffered the second formative catastrophe of his young life when his right arm was mangled by machinery he was not, as a child, supposed to be operating, and had to be amputated. Now useless for factory work, Davitt was supported by Wesleyans, who provided him with an excellent education that made allowances for his religion. Still, it is perhaps no surprise that the young Davitt – no doubt to the Wesleyans' horror – became a committed Fenian at the earliest opportunity, smuggling arms to Ireland (insert your own pun here), nor that his life continued to unspool like a Monty Python parody of a novel by Émile Zola. The Fenian movement was riddled with spies, and since a one-armed Fenian with a broad Lancashire accent was not difficult to spot, Davitt's days as a successful revolutionary were inevitably numbered. He sent his family to America for safety, but before he could join them he was arrested and jailed, serving seven years in Dartmoor Prison, where, having only one arm, he was obliged to pick oakum with his teeth. While in prison, like many inspirational figures before him, Davitt decided to write, and the numerous political treatises that resulted were all addressed to an audience of one: his pet blackbird, to which – always true to his principles – he somewhat reluctantly granted freedom, hoping it might choose to stay with him. It did not.

On his eventual release, Davitt founded the Land League and was instrumental in the strategy that became known as boycotting. The particular genius of the strategy was to harness all the bile in country life. Recognizing that small rural communities the world over have, since time immemorial, been hotbeds of jealousy, rumour, vicious enmities and spite, the Land League decided to redirect this poisonous energy at the landlord class that exploited its poor tenants. In doing so, it effectively unionized rural Ireland, channelling centuries of opprobrium towards both landlords and scabs. Davitt's league quickly created a situation in which tenants refused to pay rent (wavering tenants were ostracized), and some landlords brought in Orangemen to work the land, and soldiers to protect them. Almost overnight the entire land system in Ireland became unworkable.

At this point in his career, Davitt's efforts had all been in the service of the Irish Catholic poor, but in later life he became a boundlessly energetic internationalist, sitting as an MP in London, campaigning for the restoration of Sun Yat-sen (the first president of the Republic of China – hence the flag), resigning his parliamentary seat in protest at the Boer War, supporting Indian nationalists and reporting on the treatment of Jews in Russia. (Al was delighted to discover that some of the injustices against which Davitt campaigned were not even Britain's fault.) Davitt is that rare thing: an Irish nationalist who saw beyond both nationalism and Ireland.

After the video, Tracy treated us to a bespoke tour of the exhibition, a well-presented display of information panels interspersed with Davitt-related artefacts: first editions of his impressively eclectic publications, gifts from communities all over the world and a collection of exotic postcards that Davitt sent home to his children. Later she led us out into the grounds, where restoration work was being carried out on the thirteenth-century Straide Abbey. Though Davitt's wife is buried in Glasnevin, it was Davitt's wish to be buried as close to home as possible, and beyond the abbey lies his modest grave.

Despite the efforts of historians like Carla King, in the public consciousness Davitt remains firmly associated with two things: his own personal tragedies and the Land League campaign, and I had come to the museum expecting to be bombarded by misery. But the museum paints a far more complete (and surprisingly upbeat) picture. This is one of the great joys of small museums: what might merit one brief panel in a larger museum is given space to breathe, and the portrait of Davitt that emerged was complex and nuanced. I came away refreshed (and not just by the tea and biscuits). Even Al enjoyed it.

Death

'My hand rubbing the
Name on the headstone,
As if it were a raised scar.'
— JACKIE GORMAN,
Grave Stone

Glasnevin Cemetery ⑨

⑮
National Museum,
Collins Barracks ■   St Michan's Church ⑥ ■

River Liffey

DUBLIN   ⑤ National Museum
                         of Ireland, Kildare St ■

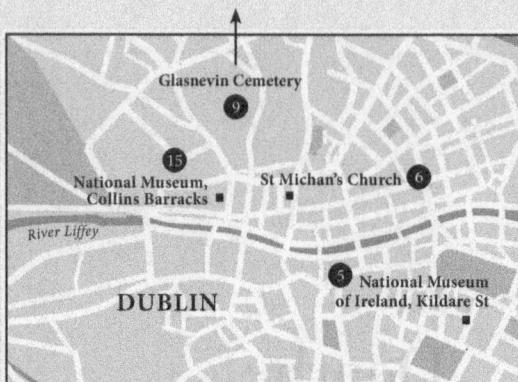

                                   Dunluce
                                   Castle ■

                    ③
                 Carrowmore
              Megalithic Cemetery
                    ■

④ Museum of Country Life
       ■

                                Tomb of Edmond Golding
                                & Elizabeth Fleming

                    ② Newgrange
         ⑩   ■
      Abbey Graveyard
① Meehambee Dolmen ■
                         Castletown
                         House   □ Dublin
             Charleville Castle ■
   ⑬ Leap Castle ■   ⑫ Hellfire Club

                     ⑧ Dunmore Cave ■ Duckett's Grove
                            ■
                      ⑦ Medieval Mile Museum

           Tomb of James Rice ■

                         Loftus Hall ■

   ⑭ Nano Nagle Place
            ■
      ⑪ Charles Fort   N

According to my aunt, Grandad Crowe was 'an awful trickster'. Sometimes his love of a practical joke extended even to death. When an aged uncle died, his body was laid out in the house and Grandad decided to add some drama to the wake. He tied one end of a piece of string to a wrist of the dead man, the other end to the door handle. Every time the door opened to let in another wave of mourners, the string jerked and the corpse's hand rose to beckon them in.

Death welcomed me in many places as I travelled around Ireland. The country is full of the dead, buried in marked and unmarked graves, in official and unofficial burial grounds, and ghosts, untethered from their final resting place, roam the country by day and night. Tokens of death, relics and memento mori can be found everywhere: in our homes, our museums, our places of worship, on our roadsides. The dead have even altered the topography of the country. Megalithic and modern tombs punctuate the landscape, while the dead in their graves slowly push the soil up so graveyards rise and churches seem to sink.

One summer evening I had dinner with my cousin Jackie in Athlone. Over dessert the conversation turned to my trip. I was mulling over what megalithic tombs to visit when she announced that there was a dolmen 'just out the road' and 'sure it wasn't going to get dark just yet'. We took a liberal view of what light is available at dusk, and off we headed. It turns out that 'just out the road' is actually a bit of a schlep and that even in high summer there's not much light at 10 p.m. under a canopy of trees. We parked in a layby and set off up an overgrown path, using the torch on my phone to light the way. Despite our best efforts to break ankles and garrotte ourselves with barbed wire (who puts up signs to a dolmen and then surrounds the dolmen itself with barbed wire?), we found it. The Meehambee Dolmen, covered with lichen and moss, squats like a benign toad in a grove of trees. The back stone has collapsed, leaving the enormous capstone (estimated to weigh twenty-four tonnes) tilting at a sharp angle towards the sky. For thousands of years this huge tomb has kept watch over the bridle path that was once a thoroughfare to the west, a silent marker of life and death and transition. There may well be human remains resting below it, but it has never been excavated, and I found something comforting in that. Sometimes a final resting place should be final, though as I discovered on my trip, that's not always the case.

More than fifteen hundred megalithic tombs have been identified in Ireland, and the most famous of these are at Brú na Bóinne, where the three Neolithic passage tombs of Newgrange,

Nowth and Dowth were constructed about five thousand years ago. As the visitor centre at Newgrange makes clear, the whole area around the bend of the River Boyne near Slane, County Meath, was thriving when the tombs were constructed. Close to both the river and the tombs there is evidence of cooking pits, sheep and cattle bones and grains of cereal, indicating that there was a very early farming community living there and using log boats to travel up and down the river. Their wooden houses are long gone, faint traces of postholes offering up hints about those who lived there, but the stone markers of death remain. The fact that remnants of death, be it bodies or grave goods, are often the only items preserved, lends a morbid pallor to the distant past. The lives our ancestors lived have evaporated, leaving only their burials behind.

Prehistoric sites display a strange reverence for the archaeologists who uncovered them. The historians who've researched and written about historic sites are never mentioned (except in the occasional book in a gift shop), but at sites like Newgrange and Carrowmore in Sligo the archaeologists take centre stage. At both sites our guides told stories about the exploits of Michael O'Kelly, George Eogan, Michael Herity and Göran Burenhult. We heard almost as much about their academic spats as we did about why the tombs were built and who constructed them. These stories are of little interest to the casual visitor, but I was fascinated, for several of these illustrious figures had taught me when I was (briefly) an archaeology student in University College Dublin. I recall one occasion when Michael Herity, a large grey-haired man with thick-rimmed glasses, swept into the lecture theatre wearing his academic gown, but instead of delivering a lecture he made us write out a hundred lines as punishment for someone whispering as he made his entrance. Soon afterwards, I dropped archaeology.

From the visitor centre, I took a bus to Newgrange with Al, Jack, Abbey and Lucy. The tomb is perched at the top of a small hill and its white quartz and granite façade glints in the sunlight. The reconstruction of the tomb is the vision of Michael O'Kelly, one of the archaeologists who worked on the site in the sixties and seventies. There were, of course, no plans or written accounts of how it looked five thousand years ago and so, based on where the white stone was found, O'Kelly re-created the exterior of the tomb as he imagined it had been (in fact, recent thinking suggests that the façade would not have looked as it does today). While Lucy turned cartwheels across the grass, Jack and I wondered at the skill and ingenuity involved in dragging the enormous boulders (up to five tonnes in weight) into position all those millennia ago. It's an engineering marvel that still isn't fully understood, but ropes, levers, wooden pulleys and huge numbers of men (and probably women) doing backbreaking work were certainly involved.

As we queued to get inside the tomb I heard Al being buttonholed by an Irish visitor who, having overheard his accent, took great pleasure in explaining very carefully and loudly to him that Newgrange was older than Stonehenge. Al was aware of this, having been to Newgrange before, but he smiled politely. Our guide appeared and delivered an introduction outside the entrance to the tomb, pointing out the famous interlocking spirals carved on the large stone at the front. What, she asked us, did we think they might mean? Abbey suggested it was a map showing Newgrange, Nowth and Dowth, while Jack thought it might be a plan of the interior of Newgrange itself. Their suggestions seemed as valid as any other I'd heard, but our guide shook her head. The spirals, she said, symbolized life, death and rebirth. How she knew this she didn't explain; our guide wasn't heavily armed with facts and tended to refer to the people who had carved the mysterious markings as the 'ancient peoples' ('Surely they must have been quite young,' quipped Jack). But she did have one fact. Newgrange, she told us, is older than Stonehenge. (Al began to groan softly.) As we shuffled down the tunnel towards the end, the children were fascinated by the letters and names carved into the stones by eighteenth- and nineteenth-century visitors, or 'vandals', as anyone who tried to emulate them would be regarded today. Even vandalism becomes respectable, given enough time. For all we know, the 'ancient peoples' who carved the symbols on the entrance stone were regarded as teenage delinquents and run out of the place.

Aside from being older than Stonehenge, Newgrange is best known for an astonishing phenomenon straight out of *Raiders of the Lost Ark*. Those who built it aligned the small lightbox at the entrance perfectly so that at dawn on the morning of the shortest day of the year the sun's beam of light would shine directly through it and illuminate the chamber inside. There's something wonderfully optimistic about an Irish monument organized around the expectation of good weather, but every year, weather permitting, a lucky few, chosen by lottery, get to witness this sight on 21 December. The rest of us make do with a re-creation of that dawn light. Our tour group huddled together in the chamber and turned to face the entrance. The lights were turned out and, as darkness enveloped us, it was easy to feel that we'd been transported back several thousand years to dawn on the winter solstice. As we gazed down the tunnel, a tiny spot of light appeared at the top of the entrance and slowly crept up the pathway until the entire chamber was flooded with golden light. I'd seen this many times before, but each time it's breathtaking. Maybe one day I'll see the real version and be able to compare.

In Ireland, enmity faces east. As you travel west, the spectre of Britain fades in the popular imagination and its role as bête noire is taken by Leinster. While at Newgrange, tour guides told us that it was older than Stonehenge; at Carrowmore Megalithic Cemetery just outside Sligo town, we were repeatedly informed that Carrowmore is older than Newgrange (Al chuckled into his hand). Carrowmore is a cemetery with thirty monuments – mostly stone circles and dolmens, with the large cairn, Listoghil, as the centrepiece. Our guide told us about Roger Walker, a local landowner who raided most of the tombs and assembled a valuable personal collection of grave goods, which he sold to the Duke of Northumberland in 1854. Some of Walker's purloined collection is now in the National Museum, but much of it remains in the Duke's ancestral seat of Alnwick Castle (now most famous as Hogwarts in the Harry Potter films). Our guide was a good storyteller and told tales associated with the saddle-back hills that surrounded us. According to local legend, the megalithic cemetery was created when the Cailleach Béara (the

Hag of Beara) flew from her home in the Ox Mountains carrying an apron of stones she wanted to use as an enclosure for her animals. But the stones fell from her apron, creating Carrowmore. As an explanation, it seems hardly more improbable than the engineering feats required. In a cairn on Knocknarea, the hill overlooking the site, stands the body of Queen Maeve, the fearsome ruler of Connaught who was killed by a hard piece of cheese fired from a slingshot, perhaps the strangest demise in Irish mythology. According to legend, she is buried upright inside the great cairn wearing full armour and clasping sword and shield, ready to do battle with her enemies. I hope that the cairn at Knocknarea remains unexcavated, just in case Queen Maeve is not inside. The story is all the better for the fact that it might be true.

Grandad Crowe, the trickster, was also a turf cutter. He cut and footed and burned turf for decades, at home in Clare and later elsewhere. When he moved east, he and a group of fellow-Claremen rented part of a bog in Wicklow, not far from Glencree. For a couple of weeks every spring he used a slean to cut hundreds of sods, which were stacked and dried in the bog before being taken home and used throughout the winter. There's nothing more evocative than the smell of a turf fire, and I have a very romantic vision of the bog and of turf cutting. In my mind's eye, I see my grandparents and their friends driving out to Wicklow armed with a picnic and a flask of tea and spending days in spring sunshine, gossiping and singing as they worked. It's a vision based largely on a photograph I have of my grandad, in a white shirt, tie and braces, cutting sods. But that romantic vision was punctured the day I took Nana to the Museum of Country Life just outside Castlebar. I thought she'd relish seeing the displays about peat cutting and harness making. Grandad had been a harness maker and I was fascinated by the tools he would have used – the stuffing stick used to fill the horse collar with rye straw, the round knife used to cut the leather, the pincers and gimlets and awls, and the huge curved needles used to sew the leather together. But as we made our way around the exhibition I noticed a lot of tutting and sighing from Nana. 'What's wrong with you?' I asked. Nana swept an arm in the direction of the display. 'Shure what would they want with that?' she said. 'Chrisht, who'd keep that sort of thing and who'd come and look at it? We couldn't wait to get rid of them.' It made me wonder: at what point does an everyday object become a valuable portal to the past? And who gets to decide? These tools had languished, forgotten and unused, in outhouses and barns, until a collector or a curator recognized their significance. But to Nana it was all junk. Nana was no sentimentalist, always looking forward (usually as far as the afterlife). She had no sense of nostalgia for a time long past. Life was hard, conditions were tough, and I didn't know I was born! I certainly didn't get my historian's instinct from her.

Perhaps there's another reason I'm drawn to the bogs, for they are natural archives, keepers of secrets; of treasures, bodies and even books. (My grandfather dug from this archive and cheerfully burned what he found. Perhaps he was an arsonist of history.) The National Museum holds many of these treasures – delicate gold bracelets and lunulae from the early Bronze Age, barrels of butter preserved for centuries, a 1,200-year-old leather-bound book of psalms, and the bodies of those buried there (often after a very violent death).

In Ireland, tourism doesn't get much darker than the *Kingship and Sacrifice* exhibition at the National Museum. The five bodies – or body parts – were all found in bogs in counties Laois, Offaly and Meath. They were all 'high-status' individuals, local chieftains or overlords, ritually sacrificed or murdered sometime between 400 BC and AD 400, maybe as offerings to the 'territorial earth goddess' Crom Dubh (a figure associated with the harvest festival), buried at the intersection of ancient boundaries and accompanied by votive offerings of the kind Nana scoffed at in the Museum of Country Life: horse harnesses, sickles, quern stones, yokes and plough parts.

Why were they killed? The local king or chief was wedded symbolically to Crom Dubh in a ceremony that reflected the dependence of Iron Age man on the land. 'It was the king's role to keep nature and society in equilibrium,' the exhibition explains. 'If the king proved himself a just ruler, then the land would respond with an increase in its fertility.' By contrast, the 'reign of an untrue king would be characterized by famine, storm, pestilence and war'. In this way, kings were hostages to fortune; in times of trouble, they took the blame. But all this is speculation, however scholarly. We don't really know why they were killed. Perhaps it was in a dispute over whether Newgrange was older than Carrowmore.

The bog bodies are exhibited in a room little bigger than a

tennis court, each one cradled by its own shell-shaped curving wall. The text describing each body is placed outside the chamber, accompanied by a pencil sketch that gives delicate visitors a warning about what they might find inside. An opening at one end leads visitors down a gentle slope, a sort of architectural drum roll, into a softly lit chamber of horror consisting of a bench built into the wall and a large glass case housing the remains.

The first time I saw the bog bodies I tutted about how they didn't really look like living people. I expected them to look like every body I'd seen laid out at a wake: serene, hands folded, with some rouge to take the pallor of death away. I wanted a Madame Tussaud's waxwork model and, rather than marvelling that anything remained of someone who had died at least 1,600 years ago, I complained that they'd let themselves go. But each return trip brings the awe and wonder that it should. This time I took Jack, Abbey and Lucy with me, and I wondered how they would react to the sight of the bodies. I thought back to my first trip to see Oliver Plunkett's head in Drogheda when I was Lucy's age and the lasting impact that had on me. By now I had taken them to cemeteries, prisons and battlefields. They were veterans of my campaign. I was confident that they wouldn't be fazed, though they'd never seen anything quite like this.

Barronstown West Man (as my nephew Jack remarked, their names all made them sound like minor heroes in the Marvel Universe) has only a rust-coloured mass of hair in the spot where his head should be. All that's left of his right hand is bone, but the left is like a twist of bog oak, with scraps of brown leathery skin, dried black strips of flesh, warped ribs wrapped around the spine. This is man as biltong. Clonycavan Man is more affecting. The horrific execution to which he was subjected sometime between 392 and 201 BC (multiple blows to the head with an axe, followed by disembowelment) was compounded centuries later by a peat-screening machine at an extraction works in County Meath, which sliced him in half at the waist. At least the second time around he didn't feel any pain. Clonycavan Man's 'extraordinary hairstyle' makes him the most hipster of corpses (beside me, I sensed thirteen-year-old Abbey nodding in approval). The back of his hair was cut short, but the top and sides had grown long enough to be gathered and tied in a knot at the top of his head, held in place with a resin imported from Spain. Perhaps the most disturbing aspect of the body is the shape of that head, no more than three inches wide and pressed flat as a Giacometti sculpture. The bog has done strange things to these men. The kids and I wondered how much he weighs. Heavy as leather, or as light as a dead mouse?

In life, Oldcroghan Man was over six foot three, an Iron (Age) giant, though it's not immediately obvious, since in death he is nothing but a headless, severed torso, the arms deflated, the hands filled out, like a Leonardo sketch in which the Old Master has focussed on one particular part. The fingers are gently curled in towards the palm, the right arm raised and the left lowered, like a cartoon of someone running. The effect is one of quiet horror. Skeletons are nothing to this. It was much easier to imagine these men alive than the skeleton of the Viking warrior we had seen in another room of the museum. Here the horror came in layers – of skin, flesh and bone, and of history too: the brutality of their deaths, the centuries-slow decomposition, the indignities of resurrection. It is not hard to imagine how some people might be overwhelmed, or why the museum takes such pains to prepare visitors for a shock. But my nieces and nephew were fascinated rather than appalled. We speculated about what the skin on the bodies might feel like – dry and papery, we decided – and they spent a long time crouched down, examining the fingers and ears, as people do with new-born babies.

If the bog bodies are treasured artefacts, displayed under glass in a room full of weaponry and jewels, they are also murder victims, an idea taken very seriously when the bodies were first exhumed. A video in one corner of the room shows Oldcroghan Man being examined by Marie Cassidy, then State Pathologist, wearing a pair of teal-coloured rubber gloves (the similarity between the hand of the pathologist and that of Oldcroghan Man's as she lifts it gently at the wrist is at once poignant and shocking). Detective Mick Macken examines Oldcroghan Man's fingers for prints. There are CT and MRI scans. I half expect them to hold a press conference announcing that they've made an arrest. I wonder what Oldcroghan Man would make of all the attention. After all, like the other victims, Oldcroghan Man was a 'high-status' person, his fingers unblemished by manual work in an era when all work was manual, and while he doubtless never imagined being turned into an exhibit, in some ways he is even more high-status now. Lying in state under glass, like Lenin, in the National M(a)us(ol)eum in the centre of the capital, he has the status of an Irish pharaoh, an ur-king. Perhaps he wouldn't complain. He might regard it as his due.

The bog bodies are not the only bodies on display in Dublin. Just behind the Four Courts stands St Michan's Church, where Handel is reputed to have practised his *Messiah* in 1742 before its premiere in nearby Fishamble Street. But few visitors are there to admire the organ or the stained glass. Instead, they come to see the crypt beneath the church where the mummified bodies have been attracting visitors since the early nineteenth century. I followed in the footsteps of those Victorian tourists. Outside the church, several iron doors sloped against the walls. They rose about a foot above the ground, and when the doors were flung open by our wise-cracking guide our small group – consisting of an American couple on honeymoon and a Norwegian woman who afterwards declared that her visit 'was the absolute best thing about my two weeks in Ireland' – descended dimly lit, uneven steps into a narrow tunnel. On each side, vaults were piled high with the dead – generations stacked on top of generations, some in elaborate coffins: faded red velvet, black leather, decorated extensively with gold crowns and brass tacks. And everything coated with a thick layer of dust. Towards the end of the tunnel was a large vault where several coffins had collapsed, and we could clearly see the bodies inside. Where the bog bodies are a leathery brown, the mummies of St Michan's are a monochrome grey, as if they have been covered by a volcanic ash cloud. They are withered, with the skin collapsing against the bones; their hair has disintegrated, but their nails remain immaculate, as if recently manicured. According to our guide, one of the bodies is known as 'The Crusader' because the way he was placed in his coffin – with arms and legs crossed – was a common burial ritual for those who had taken part in crusades to the Holy Land. Another is said to be a nun, though there doesn't seem to be any real evidence to sustain either supposition. The tour lacks detail, but I find it charming and refreshing, a relief after the intensity of the bog bodies. It reminded me a little of the Capuchin Crypt in Rome, where the bones of four thousand Capuchin friars decorate the walls and ceilings. St Michan's is less visually arresting, but equally fascinating. It may sound ghoulish, but I enjoyed being allowed into the crypt to look at the bodies and not be taught anything, for no lessons to be learned, for no science to be explained. Sometimes the experience is enough in and of itself, and for me it was sufficient to simply reflect on the fact that the same fate awaits us all.

Every new museum is a tussle for supremacy between half a dozen stakeholders, and you can always tell when the architects have got the upper hand. If, at Newgrange, the archaeologists of the sixties and seventies exert a mysterious druidic power over the exhibition and staff, at the Medieval Mile Museum in Kilkenny it appears that the designers have slaughtered the research team and made off with the entire project budget, cackling over their MacBook Pros. Opened in 2017 at the cost of €6.5 million, the Medieval Mile Museum (or MMM) aims to tell the story of medieval Kilkenny – and, by extension, Ireland. But with its whitewashed walls, glass flooring, exposed stone, minimal(ist) content and wood-panelled annexe, MMM would not look out of place on *Grand Designs*. Huge replica Celtic crosses attempt to fill the vast height of St Mary's Church, the site of the museum, the way public sculptures try to fill Tate Modern's Turbine Hall.

Like most sites dedicated to telling medieval stories, MMM is full of death. In one corner lies an open stone coffin dating from the thirteenth century. The shape of a body has been hollowed out of the stone, creating space for a corpse. The indent for the head is slightly higher than the rest of the body, so the deceased would appear to be resting on a pillow. In the base of the coffin there are holes for body fluid to drain out as the corpse decomposed. Perched on the side of the coffin was a small sign that said 'Make yourself comfortable', but I decided that's one selfie I didn't need – the idea of lying down in a coffin where human remains had drained out seemed too macabre, even for me (though I was very happy try out the [no previous owner] modern coffin that visitors can get inside at Tipperary County Museum).

I first visited MMM soon after it opened, and on my second visit eighteen months later it felt as if it was already transforming from a museum into an event space, with chairs stacked up in piles in corners and a smell of chips pervading the annexe. I had returned to see a new exhibition of three skeletons unearthed during archaeological excavations in the churchyard in 2016. Simply titled *3 Lives, 3 Deaths*, the exhibition occupies a small space in a corner of the church. It contains the skeletons, laid flat and discreetly tucked away on the far side of a square pillar, of three parishioners. One is of a woman in her late forties; the other two are adolescents. 'Concealed for centuries,' runs the blurb, 'it's time to tell their stories.' The problem, of course, is that this is largely impossible, since there's only so much that can be gleaned from bones. Their stories remain a mystery, none more so than that of a fourth skeleton – an infant – discovered during the same excavation. The panel text suggests that the child was 'abandoned or worse', a statement based on no evidence, and given that the child was buried inside the graveyard – not outside, as often happened – it seems unlikely that the child was 'abandoned'. Yet while the language about the child's death is sensationalist, the child itself is wrapped in a cloak of sensitivity, for the skeleton of the infant is not on display. In lieu of bones, a touchscreen displays images of the child's warped skull, muttering darkly about infection and physical abuse without reaching any firm conclusions. If the exhibition displays anything clearly, it's the extent to which contemporary museums are caught between the urge to display and anxiety about the ethics of doing so.

Having seen the bog bodies and skeletons, I wondered what the people who once inhabited those bodies might have thought about their remains being put on public display. In most instances, very little is known of the individual; they are simply John or Jane Doe and represent an entire community. They have no say in how they are presented. Perhaps they would hate what they've become, with eyes devouring them daily, though it's clear that some people wouldn't mind it at all: the people, for instance, who gave permission for their bodies to be displayed in Gunther von Hagens' long-running and phenomenally popular *Bodyworlds* exhibitions. Others among us are not so keen. From time to time I like to remind Al that I plan to use his kneecaps as paperweights after his death, but he doesn't seem very enthusiastic. Not that I'll be giving him a choice. He might yet thwart me by donating his body to medical science (or by outliving me), but on the whole the dead tend to lack the power of choice. They may succeed – like my grandmother – in getting the funeral they asked for, but once their descendants have forgotten them they are, by and large, fair game. You can sign all the waivers you want, but there is very little you can do – short of cremation – to prevent yourself from being dug up in a few centuries' time and displayed in a museum, stripped right down to the bone. Soft lighting and a discreet position behind a black-painted wall may add touches of veneration, of the sacred reliquary, but can't entirely disguise the reality: that even burial does not guarantee peace and quiet for ever. Whether we like it or not, we may all end up on display, objects for our descendants to study, for motives that are, very often, only partly to do with education.

Still, the display of human remains has come a long way since the Victorian era, when John Hunter, an unscrupulous Scottish surgeon, established the museum that bears his name in London's Lincoln's Inn. The Hunterian Museum (now part of the Royal College of Surgeons), which has thousands of anatomical specimens preserved in alcohol (a substance even more effective than peat) and displayed in glass jars, is a queasy mix of scientific curiosity and attraction to the macabre; one part science to three parts freak show. Its most famous exhibit used to be the skeleton of Charles Byrne, the 'Irish giant', which was acquired by Hunter after Byrne's death in 1782. The display of Byrne's body has long been controversial for, unlike most of those who find themselves in the public gaze (including the nameless bodies in Kilkenny and Dublin), Byrne *did* leave instructions as to how he wanted to be disposed of. His wish, he said, was to be buried at sea off Margate in Kent. To sink, literally, without trace. But after allegedly bribing the undertaker, Hunter thwarted Byrne's wishes and Byrne's seven-foot-seven skeleton formed the centrepiece of the collection until 2017, when the museum closed for refurbishment. But in museums around the world, attitudes to the dead are slowly changing. At present, it seems unlikely that Byrne's skeleton will return to public view, though different interest groups, primarily the Royal College of Surgeons and some academics, cannot agree on what should be done with his remains. The college may keep them, but in storage, or he may be repatriated to Ireland, or buried at sea. As a historian, I'm interested in bringing the buried past to the surface, but as a metaphor, not in reality. In the case of Byrne, his wishes should be honoured.

More human remains can be found a few miles north of the Medieval Mile Museum. Dunmore Cave is a popular tourist attraction, filled with stalactites and stalagmites, but I was lured to it not by the geology but by the story of the massacre that took place there in the tenth century, when up to a thousand locals were slaughtered by marauding Vikings (or so I thought). I took Misha along with me, who, as a veteran of the video game *Fortnite*, has slaughtered plenty of people in his time. We joined a small tour – just us, a German man, a woman from Laois, our guide and a local terrier who bounded around the cave with great surefootedness. I had read that the locals were killed when Vikings lit fires at the mouth of the cave to smoke out those who had sought refuge below. The cave filled with smoke and suffocated the people hiding there. But our guide told us that the story has now been revised. While archaeologists and scientists still think that the people were hiding from a Viking raid, they now believe that the dead suffocated, not because of Viking fires, but because the sheer numbers below ground used up all the available oxygen. Whatever version is true, it is still a horrific tale, and as we descended several hundred slippery steps into the cave I imagined the reaction of those who had discovered the bodies – all dead, but without a single sign of a blow. I wonder how they tried to make sense of this disaster. It's no wonder that the *Triads of Ireland* – tenth-century Irish manuscripts – referred to the caves as one of the darkest places in Ireland, presumably both literally and metaphorically.

At first, there's no sign of the 'massacre'. Our guide explained about the geology of the cave, and described the mysterious and beautiful sapphire and crystal lake that lies just beyond it, accessed through a complex web of tiny gaps in the limestone rock that only the most experienced pot-holers can navigate. We marvelled at the stalactites and stalagmites and were shown where a hoard of silver and bronze coins, buttons and ingots was found in 1999. And then our guide turned off all the lights and we were plunged into darkness. I could hear Misha, who found the cave claustrophobic, grab hold of the railing in front of him. We listened. The cave was alive with the sound of water dripping from the stalactites on to the cave floor and the steel platform on which we were standing. It was like being immersed in a Philip Glass piano piece. When the lights came back on and our eyes had adjusted, our guide suggested that we lean over the rail and look below. Beside a foot-high stalagmite we could make out the skull of a small child embedded in the floor of the cave. It was a moving and surreal sight. There's something heart-breaking about a body unclaimed and unburied, but in a way, having come from the Medieval Mile Museum, where bodies were unearthed and tidied up for public display, it felt less inva-sive, as if we had just happened upon a memorial and paused for a moment rather than seeking out an exhibition of bones. It seemed fitting, as we emerged in the early-evening drizzle, that there was a brief moment of sunshine which created a glorious rainbow across the fields of north Kilkenny.

I'd had my fill of looking at human remains – in skeleton or mummified form. I wanted to visit the homes of the dead, the cemeteries and graveyards. I'm particularly fascinated by old graveyards and their tombstones – their design and the inscrip-tions on them. I like to read the names and imagine the life lived by those who lie below, and I wonder about those whose burials go unmarked, for there are always many more people underfoot than there are tombstones. In Glasnevin Cemetery alone there are eight hundred thousand people buried in unmarked graves. How many stories remain unknown and untold?

The high walls and watchtowers that surround Glasnevin Cemetery make me feel as if I'm entering a walled city. And in some ways I am, for Glasnevin is a great necropolis, home to over one and a half million bodies. There are more dead people in the cemetery than there are live people in Dublin city. The walls and watchtowers were built to protect those who lay within them. In the nineteenth century, bodysnatching was a lucrative business, for not only could the bodies be sold to medical schools for dissection, but any jewellery on the corpse could be pocketed and find its way to the nearest pawnbroker. An enterprising graverobber (or 'resurrectionist', as many of them preferred to be called) didn't have to invest much in their trade. All they needed was a spade, a key-hole saw for opening the coffin, a piece of rope, a sack and a strong stomach.

The resurrectionists weren't the only ones exhuming bodies. In 1965 the remains of Roger Casement, the last of those executed in 1916, were removed from an unmarked grave in Pentonville Prison and reburied, following a state funeral, in Glasnevin. This move may have pleased the Irish government, but it would not have pleased Casement himself, since he had wanted to be buried in Murloch Bay, County Antrim – even national heroes don't always get their wish. In 2001 the remains of ten men executed during the War of Independence were exhumed from the grounds of Mountjoy Prison. Nine of them, including Kevin Barry, were reburied with full state honours feet away from Casement: the tenth, Patrick Maher, was buried in Ballylanders in Limerick, on this occasion a decision which did honour his family's wishes.

The largest memorial in Glasnevin Cemetery is that of the man regarded as its founder: Daniel O'Connell. His huge oak coffin lies in a small chapel beneath a round tower that stretches 180 feet into the air. The tower is much taller than any of the medieval round towers that inspired it – almost twice as high as the tower at Glendalough – and from the top, the entire cemetery (and most of the city of Dublin) can be seen. O'Connell is best remembered for Catholic emancipation – when Catholics were allowed to become Members of Parliament – and the Monster Meetings held to generate support for his campaign to repeal the Union, but he was active in many aspects of Irish – particularly Catholic Irish – life through much of the first half of the nineteenth century, including a campaign to establish cemeteries in Dublin where people of all faiths and none could be buried. Goldenbridge Cemetery in Inchicore opened in 1828 and Prospect (Glasnevin) Cemetery followed four years later.

When O'Connell died in Genoa in 1847 his last words allegedly were: 'my heart to Rome, my body to Ireland, my soul to heaven' – proof that the Irish continue to harbour diasporic ambitions, even after death. I've always thought it unlikely these words were uttered with his dying breath, but his heart did go to Rome and was later lost – it was placed in a silver casket and kept in Sant'Agata de' Goti church, which was associated with the Irish College in Rome, but when renovations were being carried out in the twenties it was discovered that the casket was missing, and it's never been found. Meanwhile, O'Connell's body was returned to Ireland, where, amidst great fanfare, it was placed in a simple vault in what became known as the O'Connell Circle.

The cost of plots and vaults close to the O'Connell Circle rose dramatically as the well-heeled of Dublin jostled to purchase space nearby. The cemetery is a city of the dead, and, like in any city, the price of housing is a hot topic. Most burials at Glasnevin today are in plots already occupied by a family member, but there are still some new ones available, and location, location, location dictates the price. If you want a premium spot near the entrance, close to names that litter the pages of Irish history books, then be prepared to fork out more than €30,000. If you're less bothered about your neighbours, you can purchase a plot for about €2,000. But those who'd paid a premium to lie close to 'The Liberator' saw their investment collapse when, in 1869, O'Connell's body was removed from the O'Connell Circle and placed below the Round Tower.

Several thousand burials and cremations take place every year at the cemetery, but the numbers of the dead passing through the gates is now far outstripped by the living, with over fifty thousand people taking tours every year. When I first started bringing groups of students to the cemetery, there was no museum, no gift shop, no coffee shop and very few tourists. I would phone Shane MacThomáis, then the only tour guide, and arrange to bring out a bunch of students for a visit. We'd spend hours ambling up and down the paths, pausing at the graves of the famous – Daniel O'Connell, Charles Stewart Parnell, Michael Collins, Éamon de Valera – but often spending far more time visiting the graves of the lesser known, such as Maria Higgins, the only person in the cemetery to be buried twice. She was first 'buried' in 1858 as part of an elaborate insurance scam. When rumours circulated that she was in fact alive and well, her 'body' was exhumed and, when the police opened her coffin, it was found to contain bags of sand. Following an investigation, police concluded that Maria had been coerced into agreeing to the plan, but her husband, Charles, was convicted of fraud and sentenced to two years in jail. It was another thirteen years before Maria Higgins was buried in the cemetery; this time, she really was inside the coffin.

On Shane's tours we would always stop at Parnell's grave, one of the simplest in the cemetery – a large granite stone brought from Avondale, his estate in County Wicklow. There are no religious symbols carved on the stone, no date of birth or death, just the name 'PARNELL' carved in block capitals and painted black. Shane would briefly run through Parnell's political career and the history of the Home Rule movement before pointing out that Parnell's final resting place was on top of an old cholera pit and he was buried there against the wishes of his wife and family, who had wanted him to be buried in Mount Jerome Cemetery on the south side of the city, where his father and other family members had been laid to rest. The Parnell family were overruled by a self-appointed committee who were determined that he would be buried with all the pomp and ceremony that had been afforded to O'Connell more than forty years earlier, even if his gravestone would be rather more modest.

Dressed in a tweed jacket and cardigan with a skinny tie that was often askew, Shane was charming and loquacious. He had a ready and dry wit that was sometimes lost on the students and he put his own spin on the cultural and social history of Dublin through stories of its poets, writers and singers, and the thousands of unknown and unnamed victims of cholera and other diseases that spread through the city in the nineteenth century, or the thousands of children buried in the Angels plots. His tales about pints being left on playwright Brendan Behan's headstone, and his explanations about why bodies had to leave the house feet first (to prevent the dead from looking back and inviting someone else to come with them), and coffins fitted with bells just in case the dead weren't truly dead, kept the students spellbound. We always finished the tour by drinking creamy pints of Guinness in John Kavanagh's pub (better known as The Gravediggers) on Prospect Square. Kavanagh's owes its very existence to the cemetery, for it opened to serve those visiting and working there. In fact, it served people so well that frequently coffins were left piled outside the cemetery gates when mourners failed to make it out of the pub in time to bury their dead. Within four years of the cemetery opening in 1832, a new rule came into effect: burials had to take place before midday – it was hoped that this might keep the mourners sober, at least until after the coffin had been lowered into the ground.

The mix of tour groups and mourners at Glasnevin can on occasion feel uncomfortable. It's a working cemetery with burials and cremations daily, a place where people come to mourn and reflect. But it's also a tourist destination, a place to meet for a coffee and to buy a Patrick Pearse mug, a Michael Collins coaster or a de Valera magnet. I've attended a cremation and emerged with other mourners to find not only several hearses queuing up at the front of the crematorium, each waiting to disgorge its cargo to the flames, but also large groups of tourists being corralled by tour guides busy pointing out the exquisite heads sculpted by James Pearse (Patrick Pearse's father) on the mortuary chapel or discussing the ego battles that raged between architects employed to design lavish tombs and mausoleums. Every site is subject to some degree of 'mission creep' nowadays, as they come under pressure to monetize every aspect of their operation, but perhaps cemeteries and churches are the most uncomfortable in this respect. There's no clash quite as dissonant as the one between spirituality and commerce.

When I'm not on a tour I always head for the oldest part of the cemetery, which is the most interesting in terms of the art and architecture of death: the classical columns and obelisks, the Gothic Revival memorials, the replica high crosses of the Celtic Revival. I pause by the saddened angels, the cherubs, the skull and crossbones, and the nationalist imagery of sunbursts and shamrocks, harps and wolfhounds. I'm fascinated by the way people want to be remembered (or, more likely, how those left behind think the dead should be remembered): as patriotic, as pious, as rich. Before I leave the cemetery I always visit the grave of Michael Carey, the eleven-year-old boy from Francis Street who died of tuberculosis and on 22 February 1832 became the first person to be buried in the cemetery.

Victorian cemeteries like Glasnevin are too young to feature some of the rarest and most remarkable tombstones I know: cadaver tombstones which depict the dead, not clad in their finery but as decaying corpses. These macabre tombstones are found across Europe, beginning in the fourteenth century in the aftermath of the Black Death, which decimated the population and heightened people's preoccupation with death and the afterlife. There are very few examples in Ireland, but some are easily accessible – the tomb of James Rice in Christ Church Cathedral in Waterford, Thomas Ronan in the Triskel Centre in Cork City, and that of Edmond Golding and his wife, Elizabeth Fleming, in the grounds of St Peter's Church of Ireland in Drogheda. The people commemorated in these extraordinary tombs were people of wealth and status (a little like the bog bodies in the National Museum). Rice was eleven times Mayor of Waterford, and Ronan was Mayor of Cork. In Waterford, a carved representation of Rice's body lies across the top of his tomb (his wife, Catherine Broun, is also in there, but doesn't get a carving of her own). The shroud in which he was buried is tied at the top and bottom, but open in the middle to reveal his desiccated corpse. A skeletal face with a thin layer of skin stretched across it stares up stoically at the ceiling, while the putrefying body is feasted on by small creatures, including worms, a lizard and a frog. In Drogheda, the Golding/Fleming tombstone stands upright in the boundary wall of the church graveyard. Though it stands tall now, when the tomb was made in the 1520s this slab would have been laid horizontally across the top. As with James Rice, the burial shroud has been pulled aside to reveal the decaying bodies (in both cases with far more ribs than would be anatomically correct). Each of the tombstones carries a salutary lesson – the body will be swallowed up by the land; it is the soul that must be nourished. It's a chilling visual reminder of mortality carved for a largely illiterate world: 'As you are now, so once was I. As I am, so you will be.'

*

'STOP.' The word, illuminated in bright red, flashed at me. At the same time, I saw the temperature gauge in the car shoot up to dangerous heights. I pulled over with a growing sense of despair. I was beginning to feel that the book was cursed. This was the third time I'd broken down. The first time was late on a Sunday night in November on the M6, just outside Cashel. The rain poured down, it was freezing cold and the car was constantly buffeted by passing traffic. The recovery truck arrived and loaded my inert car on to the back. I was driven to a roundabout, where I got into a taxi that took me on to Cork. My taxi driver spent most of the hour-long drive filling me in on the macabre detail of a local but very high-profile murder case. I'm all for a gruesome murder story, but hearing it while sitting in the car of a total stranger, on a pitch-black November night, was rather more unsettling than I'd like. Six months later my car once again gave up the ghost, once again near Cashel. I got another taxi to Cork and halfway through the journey discovered that my second taxi driver was the brother of the teller of murder tales. My life was beginning to feel like a sketch from *The League of Gentlemen*.

Now, sitting in my broken-down car in Athlone, I, not for the first time, thought back to that presentation in Dublin Castle where I'd been asked if I thought that visiting places associated with misery, suffering or death had any psychological impact on me. At the time, I had dismissed the suggestion, but there had been a few occasions on my travels when I empathized with Maurice Hearne in Kevin Barry's *Night Boat to Tangiers* when he complained of 'Fucking Ireland. Its smiling fiends. Its speaking rocks. Its haunted fields. Its sea memory. Its wildness and strife. Its haunt of melancholy. The way that it closes in.' Sitting in the car as dusk fell, I felt more than a 'haunt of melancholy'. Still, at least this time I wasn't stuck on the side of a motorway. I popped open the bonnet and watched as the steam escaped, like the departing soul of the engine. And then I looked around me at the place where my car had chosen to break down. To my left was Athlone Workhouse; to my right, the old Franciscan abbey graveyard. It felt somehow appropriate. Some sixty-five years earlier, my grandmother, with my father and aunt as toddlers, had regularly walked past that graveyard on her way between her mother's house and her own. Today, the graveyard is neat and tidy, with the tombstones lined up against the walls and the rest converted into a public park, but back then it was derelict, with headstones listing in all directions, graves exposed and human bones scattered across the ground. It was a place for bored teenagers to hang out, and one of their forms of entertainment was to terrify passing pedestrians. In those days, public lighting in Athlone was turned off at 10 p.m. The teenagers plucked skulls out of graves, lit candles inside them and placed them on top of tombstones. Then they waited. As Gran approached pushing a pram, my young father trotting alongside, they hid behind the gravestones, raised the illuminated skulls and bellowed, 'Goodnight Missus O'Brien!' And every time it happened my grandmother covered the half-mile home in record time, the pram bouncing in front of her, my father scampering to keep up.

Such a formative experience might explain why, as a teenager, my father regularly sent in ghost stories to the *Evening*

Herald. The column published only 'true' ghost stories (whatever those might be), and Dad's were certainly invented. Nonetheless, my father's stories were eagerly snapped up, at ten shillings and sixpence apiece. He was so successful he feared that they'd stop publishing his stories, so he submitted his tales under his classmates' names and shared the proceeds with them.

The fact that the *Herald* wanted 'true' ghost stories says something about the residual credulity of the age, but our own age is far from immune. In fact, ghosts are big business in contemporary Ireland, whether we believe in them or not. There's not a castle that hasn't a knight haunting its ramparts (*Hamlet* could have been set here), or a ghost in the dungeons rattling their chains. Cork City Gaol has hosted a ghost convention, while Spike Island offers 'after dark' tours of the fortress and prison which follow in the footsteps of paranormal investigators. And ghosts crop up in other places. At the Gothic Duckett's Grove near Carlow town, it's said that the sound of an organ being played can sometimes be heard and a ghostly foxhunt traverses the grounds, but on the day I visited with my friend Jude the only sounds we heard were the crows cawing as they flew in and out of the ruin. Another Gothic Revival castle, Charleville in County Offaly, is said to be haunted by the ghost of Harriet, the young daughter of the Earl of Charleville, who died in the 1860s when she fell as she slid down the bannisters.

On one of the hottest days of the year, Misha and I visited Charles Fort in Kinsale. We'd gone, ostensibly, to learn about the 1690 siege, but I was more interested in the fort's tragic ghost story. We headed for the ramparts, zig-zagging around pockets of lovelorn Spanish teenagers who appeared to be at a speed-dating convention. From there, we could see the port town of Kinsale to our right, while to our left families picnicked on rocks and swam alongside the sheer walls of the fort. I'd first heard about the 'White Lady of Kinsale' from Jessie. The 'White Lady' had a reputation for shoving unsuspecting people over ramparts, down stairs, and generally creating mayhem whenever she appeared in and around the fort. Jessie had recently slipped and broken her ankle near the fort and, while most rational people would attribute her fall to uneven ground or a random misstep, I was taken with the idea that it could have been the action of the malicious ghost. The 'White Lady' is Wilful Warrender, the daughter of Colonel Warrender, who was commander of Charles Fort in the late seventeenth century. Warrender was a harsh leader, a stickler for precision and unforgiving of those who disobeyed orders. His daughter fell in love with Trevor Ashurst, one of the officers at the fort, and they married. After the ceremony, as they walked along the ramparts, Wilful spotted some flowers growing on a rocky outcrop below. Her new husband persuaded the sentry to climb down and pick them; in return, he would take his place for night duty. That he did, but, exhausted after the day's celebrations, Ashurst fell asleep. It was a night when Colonel Warrender decided to inspect his nightwatchmen, and he came across the sleeping sentry. Outraged at this dereliction of duty, Warrender shot and killed him, before realizing with horror that he had murdered his new son-in-law. A distraught Wilful, still wearing her wedding dress, hurled herself from the ramparts, followed soon after by her devastated father. A day that had begun with a celebratory wedding ended with the deaths of three of the principal participants. Ever since, a ghostly woman in a wedding dress has been said to wreak havoc whenever she appears.

A different 'White Lady' haunts Dunluce Castle, perched precariously on the high cliffs of the Antrim coast. It's another tragic story, this time of thwarted love. Lord McQuillan disapproved of the man his daughter, Maeve, wished to marry and so he imprisoned her in one of the castle towers. She escaped and fled through Mermaid's Cave, the huge cavern below the castle, where the young man she was in love with was waiting for her with a boat. As they sailed to freedom, a huge storm blew up, dashing the boat against the treacherous cliffs, and the couple drowned. Maeve is said to haunt the tower, her distraught cries echoing through the castle.

As a child, I was terrified of ghost stories. The one I recall most clearly is one I heard time and again on my trips around the country – almost identical versions are associated with Loftus Hall ('Ireland's most haunted house'), Castletown House in County Kildare and the Hellfire Club (a gambling and drinking den) on Montpelier Hill in County Dublin. I heard the story first at the Hellfire Club, when I was an eleven-year-old Cub Scout sitting around a campfire not far from where the story is set. The Hellfire Club met in a hunting lodge built in 1725 for William Conolly, the richest man in Ireland and Speaker of the House of Commons. After Conolly's death, the lodge was taken over by the Hellfire Club, a group of well-to-do noblemen who drank and gambled and caroused there, men described by

Jonathan Swift as a 'brace of monsters, called Blasters, or blasphemers or bacchanalians'. There were rumours that they took part in satanic rituals, and at every meeting an empty chair was left for the Devil. Late one evening, a handsome stranger arrived to play cards. He took the chair reserved for the Devil and refused to give it up. As the night wore on, the members of the Hellfire Club became drunk and lost huge sums of money to the mysterious stranger, and as dawn broke one of them bent down to pick up a card from the floor and noticed that the stranger had a cloven hoof. Suddenly sober, he screamed and, as he did, the stranger disappeared, leaving behind only a smell of sulphur. That story kept me awake in my tent all night.

I don't believe in ghosts, but I still wasn't keen on visiting Leap Castle, the 'world's most haunted castle', on my own, so I persuaded Misha to come with me. As we approached the castle, we could see crows flying out of the top of what appeared to be a ruin. Perhaps the sat nav had brought us to the wrong place. But as we drove down the avenue I realized that parts of the castle, a fifteenth-century tower house flanked by neo-Gothic wings, were intact. I pressed the battery-operated plastic bell taped, rather incongruously, to an imposing oak door, and the door was opened by a man who seemed to be both caretaker and guide. It was a bright July morning, but we were ushered into a dark room illuminated by a large fire in one corner. A number of people were already seated there, and fifteen pairs of eyes glanced at us before immediately returning to the flames. 'I think we've stumbled into a cult meeting,' Misha whispered as we took our seats at the back.

I was hoping to be regaled with ghost stories, but instead we listened to a lengthy fireside chat that included the history of the castle, which was built by the O'Carrolls, who controlled the surrounding area. They held it until the 1640s, when it was confiscated by Cromwellian troops, who granted it to a soldier of fortune, Jonathan Darby. The local history was supplemented by a sadly inaccurate history of Ireland, and I sighed as I heard, yet again, that the Irish had been 'slaves' and then that transportation had continued up to the twenties. Eventually, I interrupted. 'What about the ghosts?' I asked. It appeared that many of them were mercenaries the O'Carrolls had hired, then turned on. After a successful battle or raid the mercenaries were invited to a celebratory banquet, where they were poisoned. This wasn't the worst of it. Other unwelcome guests were thrown into a gap in the dining-hall wall and fell several floors to the oubliette (a dungeon) at the bottom. The lucky ones landed on a spike and died soon after; the unlucky were left to starve to death and forgotten, as the word 'oubliette' suggests. According to our guide, when the oubliette was emptied, hundreds of skeletons were discovered. It's no wonder their unhappy spirits are said to stalk the castle.

After the fireside tales we were sent off to explore the castle unchaperoned. Much of it is derelict, the rest in shabby, but cosy, disrepair. There's an eclectic collection of art, books, musical instruments and religious paraphernalia. There are winding staircases that appear to lead nowhere or into tiny, purposeless rooms. Misha and I climbed the dark, uneven stairs to the derelict 'Bloody Chapel' at the top of the castle where two O'Carroll brothers had an argument and one murdered the other. Reports of bright lights emanating from the chapel have persisted ever since. Certainly, there were plenty of bright lights when we visited, since we wandered around using torches to illuminate the way. Several members of our tour group refused to go up to the chapel, preferring to stay in the Great Hall. It seemed a poor choice, as it left them hovering near the oubliette – where ghosts seemed far more likely to emerge – and close to the Minstrels' Gallery, where a terrifying haunting had occurred. In the early twentieth century, Mildred Darby had hosted séances in the castle, and she reported that after one of them, as she stood in the gallery looking down on the hall, she 'felt somebody put a hand on my shoulder . . . Thin, gaunt, shadowy . . . its face was human . . . its eyes which seemed half decomposed in black cavities stared into mine. The horrible smell . . . came up into my face, giving me a deadly nausea. It was the smell of a decomposing corpse.' I sniffed the air but detected only the faint odour of cats.

After we'd explored the castle we returned to the fireside and listened to one of the visitors talk about how she had felt the 'essence' of the place and it had filled her spirit with joy. This seemed an odd reaction to a place where so many people had apparently been murdered, but then, as Misha observed, I was not really in a position to criticize. Still, I rolled my eyes a little, until Misha leaned over and pointed out that the woman beside me had activated a ghost-hunting app on her phone and was staring at the electromagnetic field meter. I could see bars rising and falling on her screen. We decided it was time to leave.

Just minutes from the castle, we passed St Kieran's rag-bush, an old hawthorn with scraps of clothing tied to its branches. St Kieran founded a monastery nearby in the fifth century. There was (and still is) a belief that if you tied a rag to the bush and prayed for a particular intention, by the time the rag had rotted away the prayer would be answered. So strong is the belief in this bush, a road-widening plan was adjusted so that the road goes around the hawthorn. I wondered whether the hawthorn was saved to ward off the spirits that allegedly emanated from Leap Castle. Rag-trees and saints and ghosts – the (un)holy trinity of pagan, Christian and supernatural belief all bound together. The story of Ireland in one hawthorn bush.

One of my favourite 'ghost' stories has no ghost at all. When working on the development of the heritage centre at Nano Nagle Place, I read the annals kept by the Ursuline and Presentation nuns who had lived on the site. One night in September 1784, two burglars placed their ladder against the wall of the convent. As they climbed the ladder, they woke one of the nuns, who 'sallied out of her cell' just as one of the would-be thieves reached the top of the ladder. 'On looking through the window, the first object he beheld was the nun arrayed in the fascinating costume of an Ursuline night dress,' the annalist recorded, 'white veil – black band – short cloak . . .' The burglars would, in all likelihood, never have seen a nun before, but 'one glance at her was as effectual as the sight of a constable'. It's no wonder he thought he'd seen a ghost. The story of the 'Convent Goblin' spread rapidly through the city and, the convent annalist noted, the convent was never threatened again.

Though I had no ghostly visions on my tour, I did have one disconcerting experience. Late one evening, I was in the stores of the National Museum at Collins Barracks with curator Brenda Malone, looking at artefacts associated with death. As we sat there surrounded by a whole smorgasbord of them, we heard a loud metallic clatter above us. 'Oh, that must be the security guards moving the metal trays I was using earlier,' said Brenda casually. A few minutes later we heard women singing at the other end of the floor. 'That'll be the cleaners,' Brenda said. We returned to admiring Michael Collins's beautifully embroidered slippers. Later, as we were leaving, we went up to the floor with the trays. The trays hadn't been moved. As we left the building, we ran into the security guard. No, he told us, there hadn't been any cleaners on our floor that evening. There hadn't been anyone at all. Later, I was told that the ghost of a young woman haunts the building. In the eighteenth century, when it was a British military barracks, she attended a party hosted by some young officers. In the course of the evening she fell from a window and later died from her injuries, but not before telling her doctor that the officers had pushed her. No one was ever charged with her murder, and her ghost is said to walk the corridors seeking justice.

Exit through
the Gift Shop

I learned an important lesson a decade ago when I brought my nephew Jack to Dublinia, the Viking-themed museum in Dublin. At first, all went smoothly as Jack raced through every floor of the museum with tremendous enthusiasm, trying on costumes and helmets, examining everything and roaring like a proper Scandinavian warrior. I was congratulating myself on a visit well managed when we entered the gift shop. It was at this point that I made a rookie error. I'd been warned by my brother not to spend a fortune on Jack, but instead of telling him that he could have one thing from the shop – as long as it wasn't expensive – I told him he had €5 to spend. With an older child, that might have been a reasonable approach, but there was a problem: Jack was only five. He had no concept of money and could not understand the price tags on any of the items. But he was a conscientious and considerate child who took my warning seriously, and within thirty seconds he had picked out a small Viking figurine that cost €3.99. 'How much is this?' he asked. It was at this point I made my second mistake: I told him it was €3.99. 'How much do I have left to spend?' said Jack anxiously. I told him he had one euro more. 'You can pick something else,' I said, like a fool, 'as long as it's only a euro.' Many readers will have little trouble in imagining the apocalypse that ensued, but for those who haven't put themselves in this absurd situation, I will be brief: we spent the next twenty minutes going round the gift shop while Jack pointed to things and asked how much they were. There was, of course, nothing to be had for €1, but on his journey Jack soon identified some fifteen other items he would dearly love to have. Eventually, through an exhaustive (and exhausting) process of negotiation, during which Al decided to take a strong interest in the archaeology books, Jack whittled his heart's desire down to two items: the Viking figurine and a plastic axe (also €3.99). I told him he could have one or the other. Jack looked at the Viking. He looked at the axe. He looked at the Viking again. Then he sat down on the middle of the floor and burst into tears. 'It's tearing me apart!' he cried.

Despite that scarring experience, on my more recent museum visits I often started, as well as ended, at the gift shop, for a quick glance at the wares usually provides a very clear idea about who the museum or site is aimed at. At places focussed on the one-time tourist, the international visitor who is simply passing through, the shops tend to be full of generic Irish gifts: shelves of sheep in various guises (slippers, backpacks, cuddly toys); plastic leprechauns with pots of gold, ceramic cottages, tea towels with soda-bread recipes and coasters with hokum Irish sayings. In many of these shops there is no connection at all between the museum and the merchandise – it's as if an outlet of Carroll's Irish Gifts has been airlifted in and dumped at the exit. Then there are the aspirational shops, those that sell items that people might actually want to receive, rather than greeting yet another pair of sheep socks with a rictus grin. These sites, like the National Gallery in Dublin and Nano Nagle Place in Cork, make a self-conscious attempt to establish the shop as a boutique retail destination for local consumers, preferably in conjunction with a café that sells expensive but delicious home-made cakes and artisan coffee. Other sites try to combine the generic souvenir with some that relate to the site, so visitors to Spike Island can pick up plastic unicorns alongside a Spike Island-branded hat and pen, while in Crumlin Road Gaol green bobble hats and Guinness coasters sit alongside cat-o'-nine-tails and plastic truncheons.

Dark-tourism sites are places where questions of taste inevitably come to the fore. It's one thing to end your tour of the Guinness Storehouse with a commemorative pint glass etched with your name, or to emerge from Seamus Heaney HomePlace clutching a book of poetry; it's quite another to walk out of Titanic Belfast with a ship-shaped bath toy that 'even has a squeak inside for those hazardous "iceberg" moments'.

On my journey around Ireland I've drunk more cups of tea and eaten more scones in museum coffee shops than I can count, and I've collected many, many souvenirs. I love mooching around museum shops, and while I justify all my souvenir shopping as 'research', the truth is I'd probably buy the trinkets anyway. For me, they serve as an immediate, often visceral, reminder of when and where I bought them, of the day itself, rather than what the object commemorates. I keep them in my cabinet of curiosities, my own secular reliquary, where a glow-in-the-dark Mary from Knock sits happily alongside a unicorn from Spike Island and a little zip-lock bag that contains a tiny fragment of brick and a piece of paper on which is typed 'Chip off the old school wall'. On the front of the bag is a line drawing of a school and a boy sitting at a desk. The boy is Frank McCourt, and the chip, purchased from the now-closed Frank McCourt Museum in Limerick, is a reminder of his school days. But pride of place in my cabinet (at the moment, for I am very fickle) is reserved for 'Billy', a besuited Lego Orangeman, complete with bowler hat and orange sash, who hails from the Museum of Orange Heritage in Belfast. I only have one, but for the full experience I should collect a whole Orange Lodge to put on display every marching season. Although, as Doagh Famine Village taught me, in the interests of fairness and balance, I would of course have to combine it with a Lego Republican Safe House.

Our tendency to commodify the past has long been parodied and satirized by artists. In 2000, just two years after the Good Friday Agreement and years before the British Army's border observation posts were dismantled, the artist John Byrne opened his Border Interpretive Centre on the invisible line where counties Louth and Armagh meet. A blue neon sign on the roof attracted passers-by and a plaque on the wall commemorated the imaginary twinning of the Irish border with the one between North and South Korea. Inside Byrne's 'pop-up' shop visitors could purchase ceramic miniature army watchtowers, soil from the border and border sticks of rock.

In a similar vein, to mark the centenary of the 1916 Easter Rising, artist Rita Duffy opened the Souvenir Shop on North Great Georges Street in Dublin, a political, artistic and commercial venture where visitors could (literally and figuratively) pick up the satirical messages. Duffy mocked many of the clichés that have come to surround both the Easter Rising and much of Irish history. Customers could purchase Birth of a Nation Salve, Free State Jam, Carson's Ulster Marmalade and Padraig Pearse Pasta Sauce. For the green-fingered, a range of Seeds of Revolution were available, including Romantic Nationalism of the 'gone wild variety'. The flowers would, presumably, thrive in the sods of border soil Byrne had sold at his interpretive centre.

Duffy and Byrne may have been poking fun at capitalism's ham-fisted attempts to disguise itself as commemoration, but it's entirely possible that everything they sold in their shops would have passed without much critical comment, had it been found on the shelves of any Irish gift shop, or indeed in the shops of some museums. If there's profit to be made, there is almost nothing too tasteless to be sold, from the glow-in-the dark condoms for sale at Chernobyl, to a Robben Island chess set with Nelson Mandela and F. W. de Klerk as the kings. In Normandy, after visiting the beaches, visitors can pick up that great capitalist game – Monopoly – in a D-Day version where beaches are valued as real estate – Omaha deemed the most valuable, Gold the least.

Ireland has not been immune from this feast of 'commemoration' kitsch, and this was most apparent during the Easter

Rising centenary in 2016, when it was difficult to find anything that hadn't been clumsily and incongruously tagged to the blood sacrifice of the rebels. My favourite item was the chocolate bar wrapped in paper bearing the text of the 1916 Proclamation and images of all seven signatories – a bargain at €2.99. There were models of the GPO, commemorative plates and candles and socks, hip flasks, shot glasses, notepads and pens. In hardware stores, souvenir hunters could pick up a 1916 commemorative manhole cover showing the Irish Volunteer Éamon Bulfin raising the 'Irish Republic' flag on the GPO. And shopfronts got in on the action. One café blended religious and republican symbolism by placing three large statues of the Virgin Mary in the window alongside the Proclamation and images of the leaders of the Rising. One gift and clothes shop, House of Ireland, used the Proclamation as the backdrop for its window display. Models decked out in the finest Irish wool and tweed lounged in front of the Proclamation while collections of jewellery, pottery and blankets were artistically arranged in blends of green, white and orange. Had the shops been there during the Rising, they would surely have been looted. Johnston, Mooney and O'Brien, one of Ireland's oldest bakeries, clearly put out that Boland's flour mills had been requisitioned by the rebels in 1916, piggy-backed on the centenary with a large billboard advertisement that showed a beautiful golden loaf fresh from the oven above the tag-line: 'The Rising. History in the Baking'. Everywhere I looked, the blood sacrifice of the 1916 rebels was used as a kind of quality assurance, as if the core brand values of the food-and-drink industry dovetailed perfectly with the values outlined in the

Proclamation. It's not hard to imagine what the rebels – many of whom were socialists – would have made of the commercial exploitation of their fight.

Alcohol and commemoration are particularly enthusiastic collaborators. If there's a significant historic event to be commemorated, you may be certain there's a drink out there waiting to be used to toast it. If you feel the need to raise a glass to those transported from Ireland, then there's Spike Island Rum. The rum comes from Bermuda, where thousands of convicts on Spike Island were transported in the mid-nineteenth century. 19 Crimes is an Australian wine brand that uses the mugshots of several transported Fenian prisoners on its labels. The wine 'celebrates the rules they broke and the culture they built', though since most of the Fenians featured on their labels fled to the United States as soon as they could, the 'culture they built' was several thousand miles away from Australia. For the technologically minded, these mugshots can even be brought to life. Install an app and wave your phone at a bottle of Shiraz to hear James Kiely, a Fenian, speak: 'They call me the informer. Well, forgive me for caring more about myself than the cause' – an unexpected deviation from the traditional republican narrative.

To honour those who fought in 1916 you can drink 1916 Proclamation Porter, which, according to Arthurstown Brewing Company, is 'an old-school Irish stout, such as those brave men and women of Wexford would have known, brewed from our own native grain, grown in the very soil they pledged to liberate, with Chocolate and Dark Crystal malts for a velvety finish, with notes of chocolate and coffee'. It would almost bring a tear to your eye, albeit a tear that contained velvety notes of ridicule. While I've complained in these pages that my great-grandfather and those he fought alongside in Wexford in 1916 have been largely ignored, I'm not sure this is the sort of attention I was hoping for.

But my personal award for most shameless exploitation of the Rising goes to the Shelbourne Hotel. Ever since it first opened its doors in 1824 it has catered to the well-heeled visitor. All manner of the good, the not-so-good and the great have passed through its revolving doors and, during the Rising, the hotel saw no reason why the armed rebellion taking place outside should disturb its guests. On Easter Monday 1916, after opening its doors to British soldiers, who used the roof of the hotel to fire on rebel positions on St Stephen's Green, 'Afternoon Tea carried on as normal, until a stray bullet clipped the petals of a very surprised lady's bonnet.' To commemorate the centenary of the damaged hat (and, by extension, the Rising), the hotel commissioned a bespoke whiskey glass complete with bullet embedded in it. The perfect whiskey to drink in this glass is, of course, Hyde, which brought out a commemorative 1916 blend (a smoky blend, no doubt, complete with notes of cordite and the iron tang of blood). Should you prefer to drink it on the rocks, you could always add an ice cube in the shape of the *Titanic*, thereby commemorating both the Rising and the sinking of the *Titanic* in one undulating wave of pleasure.

The commodification of tragedies for profit was brought home to me particularly in the shops associated with the *Titanic*. At both the Titanic Experience in Cobh and Titanic Belfast, an enormous array of branded souvenirs is available, from Christmas-tree baubles to thimbles, pens and pencils, mugs and hip flasks, T-shirts and teddy bears. The souvenirs are further complicated by the cheerful blend of fact and fiction. I suspect that many visitors (particularly the ones pretending to be Jack and Rose) see the *Titanic* story through James Cameron's film, and the fictional tragic love story of Rose and Jack appears more real to them than the lives of 1,517 largely anonymous passengers who drowned. In both museums, visitors can buy replicas of the 'Heart of the Ocean' blue diamond necklace created for the film.

In all this, there is a real danger of making light of tragedies rather than remembering and respecting them. Of all the souvenirs I saw on my travels, the one I'm most uncomfortable with is the *Titanic* snow globe, where a model of the ship sits inside a glass sphere, for ever under water. A shake of the globe cascades glistening silver foil around the *Titanic*, like shards of ice. An almost literal memorial to the dead is marketed as a cute, festive trinket. If the demand wasn't there, the *Titanic* snow globe wouldn't be for sale, but that doesn't mean it should be. I'm just grateful that a trip to a famine museum is not accompanied by the option to purchase seed potatoes or tins of Soyer's Famine Soup. At least, not yet.

As I write this, I am surrounded by the faint smell of burning turf, turf I bought in a museum shop. Tiny pieces (each one smaller than a €1 coin) are packed in a paper box shaped to look like a thatched cottage. Inside is a small stone plate on which to place and burn the pinch of turf (for that's really all it is). The smell promises to bring back memories of turf fires and a carefree past. A promise to give the gift of home. And it does smell like a turf fire (very briefly), but I'm slightly mortified as I burn it, thinking of my grandad cutting turf, but more of Nana, who would doubtless tut and exclaim that it was 'nothing but some aul codology' and I'd been fooled into throwing my money away. And perhaps she'd be right, but while it's embarrassing to have purchased this, I enjoy the very clear vision of her that it conjures up. I can hear her mocking me as plainly as if she were sitting across from me in front of the fire in her own back room. I put the pen down, close my cabinet of curiosity, turn out the light and walk away to put the kettle on.

Your Feedback is
Important to Us

In 1805, the educator Joseph Lancaster wrote a letter to John Foster, the Chancellor of the Exchequer in Ireland: 'The feelings of the Irish Nation are strong,' he wrote, 'and their passions sometimes dangerous in the extreme. It is by informing the minds ... of the people that Ireland will attain its proper dignity.' Like other members of the British establishment, Lancaster believed that a population could be taught to be loyal and responsible subjects and that, in Ireland, where the loyalty of the population was suspect, it was especially important to impart the right sort of education. Inspired by Lancaster, Dublin businessmen, including Samuel Bewley and William Guinness, established the Society for the Education of the Poor in Ireland (better known as the Kildare Place Society) in 1811. In 1831, a national school system was established, complete with textbooks that put the Empire at the heart of lessons. The *Second Reading Book*, aimed at children aged seven, informed pupils that 'on the east of Ireland is England, where the Queen lives.

Many people who live in Ireland were born in England. We speak the same language and are called one nation.' By repeating it often, maybe enough people would come to believe it. In this context, it's no surprise that museums were regarded in the same way: as instruments of establishing social order (some might argue that they still are).

Museums and historic sites have never been neutral spaces. In looking at the past – whether in a classroom, a history book or a museum – we remember what we are told to remember. As sociologist and historian James W. Loewen observed in *Lies Across America*: 'Most historic sites don't just tell stories about the past; they also tell visitors what to think about the stories they tell.' Indeed, in the late nineteenth century the National Museum in Dublin was established as a place to reflect on the spoils of empire, an empire to which many people in Ireland did not want to belong. While museums and heritage sites in Ireland are no longer showcases for propaganda about the value of the British Empire, many now tell a narrative about our past that is heavily influenced by the Irish nationalist version developed in the mid-nineteenth century. It's a version that, much like the cartoons of the nineteenth century, regards Ireland as an 'injured lady', a version in which the indigenous population plays the roles of victim, innocent and martyr. Many exhibitions build their narrative around these assumptions, and the belief in our own innocence is nowhere more apparent than at former sites of incarceration, where everyone, it seems, was either a rebel, a child or a lovable rogue. But it's not healthy for a nation to believe in its own innocence. It creates blind spots. The ironing out of history to allow for a simple narrative, whether it's to ensure innocence, highlight victimhood or create heroes, is dangerous. In Brian Friel's *Making History*, Peter Lombard, Archbishop of Armagh, insists, 'People think they just want to know the "facts"; they think they believe in some sort of empirical truth, but what they really want is a story.' And while playwrights, novelists, poets, artists and songwriters can finesse the truth in favour of the story, historians, teachers, researchers and curators have an obligation to make complex history accurate, accessible and engaging. The rise of the far right across the world has been fuelled and bolstered by an obsessive focus on perceived historical injuries suffered at the hands of other nations and ethnicities.

To some extent, museums are about how we view ourselves or, more accurately, how we would like to be viewed. They are window displays offering a range of goods, showcasing the shiniest, most dramatic, most alluring, the ones that entice visitors in. And we all put on a public face and fabricate and tweak elements of our past and our present. Facebook, Twitter and Instagram all encourage us to curate our own lives, and we do so with gusto, exaggerating, embellishing and conflating some stories while ignoring others altogether. We all favour a neat, flattering version of ourselves, so it's hardly surprising that in the creation of a national narrative the same thing happens. But this island's past is as complicated and contradictory as our real lives. Our museums should reflect that complexity.

In the middle of writing this book, a cousin of Nana's died. Bridie, or 'The Bird', as we called her, was sparrow-like: small, thin, a little beaky, always with a hairnet pulled over her tightly permed hair, a cardigan which she clasped firmly around her, a neat pastel blouse with a round collar, a straight skirt that ended mid-calf, and sensible shoes. The Bird's lack of interest in her appearance always troubled Nana. 'She could have made a lot more of herself,' she'd tut. The Bird lived a few miles from Nana and came to dinner every couple of weeks. As a child, and for long afterwards, all I really knew of her was that in the late fifties, when she was in her twenties, having never even been as far as Ennis, which was thirty kilometres from her home, she packed a bag and ran away. She had £100 in her pocket (the first money she ever had, given to her by her father after he'd sold a herd of cattle) and she intended to start a new life in England. But she never got beyond Dublin. After getting off the train at Heuston, she travelled across the city to Phibsborough, where she got a job as a housekeeper to the Vincentian Fathers at St Peter's Church. She devoted more than forty years of her life to them before retiring to a small flat, where, as far as I knew, she spent her days praying and counting her pennies (according to Nana, she was famously tight with money). As a teenager, she symbolized everything I hoped not to be. I thought her life was dull and monotonous, and I was especially infuriated by her bungled 'great escape'. Instead of having adventures, I thought, she'd wasted her life looking after a bunch of old men and their church.

Then she died, and my aunt gave me a parcel of papers recovered from her flat. Inside the parcel were some newspaper clippings – mostly songs, poems and articles about the War of Independence – and a batch of letters tied up with red ribbon. The letters, written between 1957 and 1962, were from two Claremen. Both declared their love for Bridie and hoped that a match might be made. One declared that he had 'never met anyone to compare to you' and he knew he 'could always be happy' if she would marry him; the other imagined them 'married and living happily ever after'. I only have their letters, but it's clear that Bridie did respond (fitfully), that she saw the men on her very occasional visits home and sent photos of herself. While writing to the would-be suitors in Clare, she was also dating in Dublin – at one point stepping out with a former Garda – and she had signed up to a dating agency, for there's a letter from a Mayo man living near Northampton responding to one she had written to him. Clearly, Bridie had plenty of opportunities for romance, yet she never married. And I wonder whether the clue to this mystery lies among that bundle of papers where I found several neatly folded, yellowing pages published by the British Institute for Practical Psychology. The pages outline an 'auto-psychology' course for people who had diagnosed themselves as having an 'inferiority complex' (a course they claimed was 'particularly fascinating to women of all classes'). There's no evidence that Bridie ever signed up to the course (she had not completed the enrolment form), but she had clearly thought about it, and I found that glimpse of vulnerability unexpectedly touching. The Bridie I knew was a nervy, virtually silent old woman. But the letters, the newspaper clippings and the unsent enrolment form offered tiny snapshots of her emotional life in the fifties and sixties, when she was in her twenties and early thirties. She'd led a life that was much more complex and interesting than my experience of her suggested. And although I hadn't set out to research her, it reminded me of the importance of research, and of not taking things at face value. Historians, researchers, curators and archivists all play crucial (but often hidden) roles in helping to weave together the complex, colourful tapestry of our past. Without them, we end up with a monochrome version which perpetuates tired old tropes and clichés. Some museums and heritage sites fall into that trap, using artefacts not to think deeply about the richness and complexity of our past, but to illustrate stale ideas.

I agree with journalist and commentator Fintan O'Toole's assertion that there 'is no aspect of our heritage that is not available for exploitation'. It seems that efforts to develop new stories and explore the diversity of the island's heritage are often stymied by a countervailing tendency to promote its past in a sterile way. Fáilte Ireland, the National Tourism Development Agency, wants sites to shoehorn stories into their predetermined themes. The agency is focussed on marketing rather than history and heritage, and this wouldn't matter so much if Fáilte Ireland were responsible solely for promoting Ireland abroad – after all, it's hard to sell complexity and nuance to an international audience that may know little of Ireland to begin with – but the organization also controls a sizeable proportion of the funding available for heritage development in Ireland. The agency's dominance stifles imaginative projects across the country, encouraging the same neat stories to be told at a multitude of sites. Instead, museums and heritage sites should be encouraged (and funded) to explore their own stories, to tell multifaceted tales, to engage with complex histories and to prioritize the research and the content over the desire to fit everything into tidy, branded bundles. A separate organization, one with knowledge and understanding of Ireland's history and of museums and heritage sites, should be in charge of heritage development and should work alongside Fáilte Ireland in developing a marketing strategy. Expert research, professional conservation and nuanced and engaging storytelling must be prioritized over glib slogans and high-tech gimmickry.

In this book, I've told many stories associated with the sites I visited, but you won't always discover those stories if you visit these sites yourself. This is particularly true of sites that don't have a visitor centre and rely on a handful of outdoor panels to tell their tale. Many of these are managed by the Office of Public Works (OPW) but, as its fussy bureaucratic name implies, it's not widely celebrated for its facility in bringing history to life. It's as if the memory of the nation has been entrusted to a handful of people convinced that no battle, siege or massacre could possibly be as exciting as the sight of a limestone corbel or the ornate detail on a stone boss. An information panel should encourage a visitor to learn more, or at least provide them with a memorable story. We get a much stronger sense of the past from stories than we do from sterile dates. To be fair to the OPW, they are not the only organization guilty of producing lacklustre information panels (and they do run some excellent sites, including Kilmainham Gaol, Carrowmore Megalithic Cemetery, Cahir Castle and Dunmore Cave), but it's a shame that so many of our lesser-known sites seem doomed to remain obscure when the information displayed there is so dull.

*

Where stories of war, rebellion, politics and religion are told, these are neatly divided into two sides – one good, one bad – but that simplistic narrative often requires quite a lot of fabrication and creativity to sustain. Neither the Earl of Desmond, nor the defenders of Drogheda from Cromwellian troops, nor King James and his supporters, were great Irish patriots fighting to secure a free and united Ireland. It doesn't matter how many illustrations depict Sergeant Custume in a green uniform defending Athlone from 'invaders', he remains a royalist, loyal to an English monarch.

Lives and politics and wars were (and are) complex, contradictory and confusing, and simplifying the stories, making them binary – a matter of 'them' versus 'us' – is the easy (and often the deliberate) way out. On my travels, the 'them' was almost always the English (even though 'British' would often have been more – if not always completely – accurate), the 'us' the Irish. I heard a lot about eight hundred years of oppression, about 'English' invasions, land confiscations and massacres. Many of these stories are true, but other stories could also be told. And when it came to other invaders – the Vikings, for example – a very different narrative emerged. The Vikings, insofar as they appear in our museums, are rather swashbuckling figures and, apart from the occasional critical remark about poor monks being attacked, there is no trace of any bitterness. The British Crown and its governments may have had no business being in Ireland, but English, Welsh and Scots people have been settling peacefully in Ireland since before St Patrick. Some hundred thousand British citizens currently reside in the Republic, and many Irish people bear Anglo-Norman and English surnames. The connections between the two – or rather the four – countries run very deep. But too often the stories of our shared past that appear in our museums can leave visitors thinking that there are only negative stories to be told about the relationship between the two islands.

Our museums and heritage sites face difficult decisions about inclusion and exclusion, but their solutions, for the most part, are looking increasingly out of date. There are many voices that have historically been silenced in Ireland, many stories that have not yet been told. While women tend to be included in most recent exhibitions, often they still appear as an appendage – a sister, wife or mother of a 'great man'. Recently, in a series of temporary exhibitions, EPIC has focussed on telling the stories of some women, such as the astronomer Agnes Clerke and the designer Eileen Gray, and Cavan County Museum highlights a number of women, including social justice campaigner Sr Majella McCarron and librarian Letitia Dunbar-Harrison in their *Women of Influence* exhibition. The National Museum at Collins Barracks hosted *(A)dressing Our Hidden Truths*, an exhibition of work by the artist Alison Lowry which explored the distressing and bleak history of industrial schools, mother-and-baby homes and Magdalene laundries in Ireland. It was an important move by the museum – the first national institution to directly engage with the legacy of these institutions through an exhibition – but much more needs to be done.

Far more also needs to be done both to welcome a diverse range of visitors, particularly from marginalized communities, into our museums, and to tell their stories. The prominence of a handful of iconic gay figures in Irish history, such as Oscar Wilde and Roger Casement, together with a growing recognition of lesbians in Irish history, including Dr Kathleen Lynn and Madeleine ffrench-Mullen, both members of the Irish Citizen Army who fought in the 1916 Rising and co-founders of St Ultan's Children's Hospital, has afforded the LGBTI+ community an increasing presence in Irish museums. However, for the most part, their stories are retrofitted to existing exhibitions, as with the Rainbow Revolution Trail that can be followed through the galleries at the National Museum. Travellers barely merit a mention – I only recall seeing them represented at three locations: Cork Public Museum, Doagh Famine Village and a recent temporary exhibition at the Museum of Country Life in Castlebar. Children are certainly catered for as a museum audience, with plenty of worksheets to complete and costumes to try on, but stories about children or objects associated with them are generally found only at prison sites (one exception is 14 Henrietta Street, where part of the tour includes a moving film and soundscape about childhood rhymes and games). People with physical or intellectual disabilities are almost entirely absent from the historical narrative.

We are a nation of immigrants. There have been successive waves of immigrants for centuries – missionaries, Vikings, Anglo-Normans, Old English, New English, Scots, Welsh, Jews, Muslims and thousands of others who don't fit neat definitions. Where is the EPIC for them? Our museums and heritage sites should reflect the immigrant experience as well as the emigrant.

The 'New Irish' – if they visit our museums at all – can have little idea that they are following in the steps of people who arrived many years before. And for emigrants it seems that those who left only count if they left generations ago, and really only register in broad brushstrokes of tragedy or in pen portraits of the great success stories. Nothing else matters. Much of this appears to be the result of a tourism strategy designed to chase American dollars – a curious approach, given that North Americans accounted for only 2.1 million of the 10.6 million overseas visitors to Ireland in 2018. Three and three-quarter million visitors came from Britain, and yet nothing in museums (at least in the Republic) is aimed specifically at them (except, in the case of the English, for the occasional jibe).

More than ever, museums should also reflect the present. In 2018, Brenda Malone began what she has termed 'collecting the now', installing a small display in the National Museum at Collins Barracks in Dublin which reflects both the Marriage Equality and the Repeal the Eighth referendums that took place in 2015 and 2018. Contemporary collecting is a relatively new practice in major museums but, as Brenda has observed, it offers a 'real opportunity for Irish people to say, "This is what we want remembered"', and it helps make the National Museum relevant to today. This is vital, for there is a danger that if museums don't reflect both the 'then' and the 'now', we will believe that we are, in the words of academic and literary critic Declan Kiberd, 'history's fulfilment' – that everything that has gone before was leading up to this moment – not that we are part of something that is constantly evolving. Museums and heritage sites can show us that we 'exist in time and that the world has many more experiences to offer than a depthless present'. A nuanced understanding of the past ensures that we do not accept the present with passivity. Our museums and heritage sites should hold up a broken mirror to the past, one in which we see many splintered reflections. As President Michael D. Higgins observed in a speech at the 2018 Dublin Festival of History:

> We cannot, nor should we, demand that another adhere to our own interpretation of our past. We can, however, and we must, require of ourselves and others, a transparency of purpose and honesty of intent, a serious engagement with historical scholarship and, above all, respect for the sincerely held beliefs and ideas of others, including those who went before.

When my tour was over, I put the finishing touches to the map of my trip, hung it on my wall and stood back. What the map shows, I realized, is that our museums and historic sites are far from evenly spread. If Ireland needs more museums, they should be placed anywhere other than Dublin, Cork or Belfast. There's a strong case to be made for having a museum about death and funerals. So many stories could be told about the rituals surrounding death and burial, about superstition and faith, about mourning, about the science of death and the architecture of the afterlife. But I'd also like to see a museum of untold stories, a museum that's a voice for those who have been largely silenced by history, those whose opinions didn't officially count, those forced into the margins. I'd like it to tell the stories of immigrants, of women, of non-famine emigration, of Travellers, of everyone whose stories have been overlooked or ignored. I'd like it to be full of contradictions and complexity, telling stories in engaging and thought-provoking ways. I'd like it to have a whole gallery permanently dedicated to happiness (a decent bookshop and a café with excellent home-baked scones wouldn't go amiss either). And I'd like it to reflect the present, as well as the past, for, as Mrs Boyle pleads in Seán O'Casey's *Juno and the Paycock*: 'it's nearly time we had a little less respect for the dead, an' a little more regard for the living'.

An Unexpected
Darkness

In July 2020 I found myself staring at the interior of an empty church. I wasn't in the church, I was in Liverpool, sitting in front of my laptop. After several minutes my grandmother's coffin was wheeled in and I saw the backs of my parents, siblings, uncle and aunts as they shuffled into separate pews. Two metres apart, but together. My nephew Jack WhatsApped me from the burial and, after watching the coffin being lowered into the ground, he held the phone up and turned it in a circle so I could say hello to the few mourners gathered around the grave. The call ended and I sat silently on my sofa, desperately craving all the things that ought to follow the internment: the rustle of paper being lifted off sandwiches, the whistle of a kettle, the chat, the laughter and tears. Wakes and funerals are about honouring the dead but, really, they have always been for the living.

In *My Father's Wake* Kevin Toolis observed, 'All of us need to find a way to handle death. It will be a lot easier if we just copy what the Irish already do.' But for almost two years there was no possibility of 'copying of the Irish way'. There were no wakes, no coming together, no clasping of hands, no 'sorry for your trouble'. There were no handfuls of soil being thrown on coffins by friends and family, no rosaries recited in comforting harmony through the graveyard. There were no ham-and-coleslaw sandwiches, no cocktail sausages laid out on sticky oilcloth-covered trestle tables, no jugs of diluted orange or pints of Guinness. There were no group hugs, no gales of laughter, no 'Parting Glass', no 'Good night and joy be with you all' to send us on our way as we drained our glasses and gathered our coats.

In death, everyone is alone. It's not this single moment of solitude that threw our whole way of grieving out of kilter, it's the fact that thousands of people died without family or friends with them in their final months, weeks, days and hours; it's the fact that funerals took place with only ten people in attendance, ten people who could not hug and comfort each other. Across Ireland, local communities lined the roads from homes to churches to cemeteries, two metres apart, to pay homage to those who had died. Friends, families and colleagues living further than five kilometres away had to rely on the live-streaming of the funeral service as they sat bereft at home. Slowly the numbers allowed to attend funerals increased, from twenty-five to fifty, then to half the capacity of the church or crematorium. Finally, after almost two years, on 28 February 2022, funerals took place where mourners did not have to wear face coverings and could shake hands and hug each other. What had once been small, simple gestures now seemed hugely significant.

Throughout the pandemic there was no way to avoid

comparisons with other dark days of the past – the flu pandemic of 1918, the Famine of the 1840s. During the Famine, more people died of disease than starvation and, as with COVID-19, there was great fear of contagion. And so many suffered and died alone, died in such numbers that communities were overwhelmed; there were funerals with no mourners, burials with no coffins and, in some cases, mass graves. There was no dignity in death. We are, thankfully, not living in famine times, but the suffering of individuals and of society wrought by COVID-19 was immense. Each death was a tragedy for families and friends, and having no way to publicly grieve heaped sorrow upon sorrow. Now, at last, that is possible again: we can hold back-dated memorials and celebrations, but the pandemic was a reminder of how much mourning is a collective and community activity.

Death, burial and mourning are steeped in generations of tradition. Many of those traditions are explored in our museums and heritage centres, many more are shared among family and friends around a kitchen table, a death bed, on bar stools. As we return to the full embrace of the communities that gather when a loved one dies – the community of the one who died and the communities of those left behind – there is an opportunity to take stock, to consider adapting old traditions and adopting new ones. If museums and heritage sites need to find new ways to tell old stories (and new stories to tell), perhaps we as communities need to consider new ways of dealing with death. This needn't mean rejecting the past, simply opening ourselves to the possibility that a wider variety of traditions would be more representative of an increasingly diverse and multi-cultural society.

SELECT BIBLIOGRAPHY

NOTE ON SOURCES

As a historian, I'm used to burrowing in archives and ploughing through academic texts, but for this book I wanted to use as wide a range of sources as possible. In the course of my research I visited over two hundred sites, listened to a lot of music, watched a lot of films, looked at a lot of art and talked to as many people as I could, including academics, curators, site managers, friends, family and strangers. I also read widely, not just history books and panel text, but fiction, travel writing, plays, poetry, biographies, memoirs and anything I thought would help me to get a sense of why, for the Irish, the darkness of the past still echoes so loudly.

As the name implies, a select bibliography is far from comprehensive. In writing *The Darkness Echoing* I have drawn on the hundreds, if not thousands, of books that I've read over the last two decades, many of which helped to form my understanding of Ireland's past. The books listed below are the ones I've found particularly useful, books I've returned to time and again over the course of my research. I've also listed books that I mention or quote in the text. The list is in alphabetical order by author (and I have made no distinction between fact and fiction), but if this book has sparked a renewed interest in Irish history, as I hope it has, there are two engaging, well-written histories of Ireland you should start with: Thomas Bartlett's *Ireland: A History* (Cambridge University Press, 2010) and John Gibney's *A Short History of Ireland* (Yale University Press, 2017). For reading on more specific topics, the selection below will be useful.

Barry, Kevin, *Night Train to Tangiers* (Canongate, 2019)

Beiner, Guy, *Forgetful Remembrance: Social Forgetting and Vernacular Historiography of a Rebellion in Ulster* (Oxford University Press, 2018)

Billings, Mattieu W. and Farrell, Sean, *The Irish in Illinois* (Southern Illinois University Press, 2020)

Boland, Eavan, *Outside History: Selected Poems 1980–1990* (Carcanet Press, 1991)

Böll, Heinrich, *Irish Journal* (English translation by Leila Vennewitz, Melville House Press, 1967)

Bourke, Marie, *The Story of Irish Museums, 1790–2000: Culture, Identity and Education* (Cork University Press, 2011)

Bourke, Richard and McBride, Ian (eds.), *The Princeton History of Modern Ireland* (Princeton University Press, 2016)

Carey, Tim, *Mountjoy: The Story of a Prison* (Collins Press, 2000)

Casey, Christine, *The Buildings of Ireland: Dublin* (Yale University Press, 2005)

Chambers, Anne, *Grace O'Malley: The Biography of Ireland's Pirate Queen* (Gill Books, 5th edn, revised, 2009)

Coleman, Marie, *The Irish Revolution 1916–1923* (Routledge, 2013)

Coogan, Tim Pat, *The Famine Plot: England's Role in Ireland's Greatest Tragedy* (Palgrave Macmillan, 2012)

Cooke, Pat, 'History, Materiality and the Myth of 1916', in Godson, Lisa and Brück, Joanna (eds.), *Making 1916* (Liverpool University Press, 2015), 203–17

Corporaal, Marguérite, Cusack, Christopher and Janssen, Lindsay (eds.), *Recollecting Hunger: An Anthology* (Irish Academic Press, 2012)

Crowley, John, Murphy, Michael and Smith, William J. (eds.), *Atlas of the Great Irish Famine* (Cork University Press, 2012)

Crowley, John, Ó Drisceoil, Donal and Murphy, Mike (eds.), *Atlas of the Irish Revolution* (Cork University Press, 2017)

Dickson, David, *Arctic Ireland: The Extraordinary Story of the Great Frost and Forgotten Famine of 1740–41* (White Row Press, 1997)

Dolan, Anne, *Commemorating the Irish Civil War: History and Memory, 1923–2000* (Cambridge University Press, 2003)

Dolan, Anne and Murphy, William, *Michael Collins: The Man and the Revolution* (Collins Press, 2018)

Donnelly, James S., *The Great Irish Potato Famine* (Sutton Press, 2001)

Doughty, Caitlin, *From Here to Eternity: Travelling the World to Find the Good Death* (Weidenfeld and Nicolson, 2017)

Doyle, Lynn, *The Spirit of Ireland* (Batsford, 1935)

Duffy, Joe, *Children of the Rising: The Untold Stories of the Young Lives Lost during Easter 1916* (Hachette, 2015)

Dunne, Tom, 'The Marriage of Strongbow and Aoife: Entertaining History in the Interests of the State', in Rooney, Brendan (ed.), *Creating History: Stories of Ireland in Art* (Irish Academic Press, 2016), 134–53

Elliott, Marianne, *Wolfe Tone: Prophet of Irish Independence* (Yale University Press, 1989)

Evans, E. Estyn, *Irish Folk Ways* (Routledge, 1957)

Ferriter, Diarmaid, *On the Edge: Ireland's Off-shore Irelands: A Modern History* (Profile Books, 2018)

Fitzpatrick, David, *Oceans of Consolation: Personal Accounts of Irish Emigration to Australia* (Cornell University Press, 1994)

Foley, Michael, *Death in Every Paragraph: Journalism and the Great Irish Famine* (Quinnipiac University Press, 2015)

Foster, R. F., *Paddy and Mr Punch. Connections in Irish and British History* (Allan Lane, 1993)

Friel, Brian, *Translations* (Faber and Faber, 1980)

____, *Making History* (Faber and Faber, 1988)

Gleason, A. B., 'Hurling in Medieval Ireland', in Cronin, Mike, Murphy, William, and Rouse, Paul (eds.), *The Gaelic Athletic Association, 1884–2009* (Irish Academic Press, 2009), 1–13

Goarzin, Anne, 'Articulating Trauma', *Études irlandaises*, vol. 36, no. 1 (2011), 11–22.

Godson, Lisa, and Brück, Joanna (eds.), *Making 1916: Material and Visual Culture of the 1916 Rising* (Liverpool University Press, 2015)

Gorman, Jackie, *The Wounded Stork* (Onslaught Press, 2019)

Gray, Peter, *The Irish Famine* (Thames and Hudson, 1995)

____, *Famine, Land and Politics* (Irish Academic Press, 2001)

Griffith, Lisa Marie and Wallace, Ciarán (eds.), *Grave Matters: Death and Dying in Dublin, 1500 to the Present* (Four Courts Press, 2016)

Hall, Anna Maria and Hall, Samuel Carter, *Hall's Ireland: Mr & Mrs Hall's Tour of 1840*, Michael Scott (ed.) (Sphere, 1984)

Handler, Jerome S. and Reilly, Matthew C., 'Contesting "White Slavery" in the Carribbean: Enslaved Africans and European Indentured Servants in Seventeenth-Century Barbados', *New West Indian Guide/Nieuwe West-Indische Gids*, vol. 91, no. 1 (2017), 30–55

Heaney, Marie, *Over Nine Waves: A Book of Irish Legends* (Faber and Faber, 1994)

Heaney, Seamus, *Door into the Dark* (Faber and Faber, 1969)

_____, *The Spirit Level* (Faber and Faber, 1996)

Hogan, Liam, McAtackney, Laura and Reilly, Matthew C., 'Why We Need to Confront the "Irish Slave Myth" and How Terminology is not Simply Semantics', *History Ireland*, vol. 24, no. 2 (2016), 18–22

Hooper, Glenn (ed.), *Heritage and Tourism in Britain and Ireland* (Palgrave, 2016)

Horne, John (ed.), *Our War: Ireland and the Great War* (Royal Irish Academy, 2008)

Jeffrey, Keith, *Ireland and the Great War* (Cambridge University Press, 2000)

Joye, Lar and Malone, Brenda, 'Displaying the Nation: The 1916 Exhibition at the National Museum of Ireland, 1932–1991', in Godson, Lisa and Brück, Joanna (eds.), *Making 1916: Material and Visual Culture of the 1916 Rising* (Liverpool University Press, 2015), 180–94

Kelleher, Margaret, *The Feminization of Famine: Expressions of the Inexpressible* (Cork University Press, 1997)

Kelly, James (ed.), *Cambridge History of Ireland, Volume 3, 1730–1880* (Cambridge University Press, 2018)

Select Bibliography

Kelly, James, Ohlmeyer, Jane and Smith, Brendan (eds.), *Cambridge History of Ireland*, four vols. (Cambridge University Press, 2018)

Kelly, Niamh Ann, *Ultimate Witnesses: The Visual Culture of Death, Burial and Mourning in Famine Ireland* (Quinnipiac University Press, 2017)

Kenna, Shane, *Jeremiah O'Donovan Rossa: Unrepentant Fenian* (Merrion Press, 2015)

Kennedy, Liam, *Unhappy the Land: The Most Oppressed People Ever, the Irish?* (Merrion Press, 2015)

Kinealy, Christine, *Charity and the Great Hunger in Ireland* (Bloomsbury, 2013)

____, 'Private Donations to Ireland during An Gorta Mór', *Seanchas Ardmhacha: Journal of the Armagh Diocesan Historical Society*, vol. 17, no. 2 (1998), 109–20

King, Carla, *Michael Davitt: After the Land League, 1882–1906* (University College Dublin Press, 2016)

Loewen, James W., *Lies Across America: What Our Historic Sites Get Wrong* (Touchstone, 1999)

MacSuibhne, Breandán, *The End of Outrage: Post-Famine Adjustment in Rural Ireland* (Oxford University Press, 2017)

____, *Subjects Lacking Words? The Gray Zone of the Great Famine* (Quinnipiac University Press, 2017)

MacThomáis, Shane, *Glasnevin: Ireland's Necropolis* (Glasnevin Trust, 2010)

Mark-FitzGerald, Emily, *Commemorating the Irish Famine: Memory and the Monument* (Liverpool University Press, 2003)

McAtackney, Laura, 'Graffiti Revelations and the Changing Meanings of Kilmainham Gaol in (Post)Colonial Ireland', in *International Journal of Historical Archaeology*, 20 (2016), 492–505

McBride, Ian (ed.), *History and Memory in Modern Ireland* (Cambridge University Press, 2001)

McCarthy, Cal and Ó Donnabhain, Barra, *Too Beautiful for Thieves and Pickpockets: A History of the Victorian Prison on Spike Island* (Cork County Library and Arts Service, 2016)

McCarthy, Cal and Todd, Kevin, *The Wreck of the Neva* (Mercier Press, 2013)

Miller, Kerby, *Ireland and Irish America: Culture, Class and Transatlantic Migration* (Field Day Publications, 2008)

Moloney, Mick, *Across the Western Ocean: Songs of Leaving and Arriving* (Quinnipiac University Press, 2016)

Morton, H. V., *In Search of Ireland* (first published 1930; Methuen Publishing, 2000)

Murphy, Richard, *Sailing to an Island* (Faber and Faber, 1963)

O'Brien, Gillian, *Blood Runs Green: The Murder that Transfixed Gilded Age Chicago* (Chicago University Press, 2015)

_____, 'The 1825–6 Commissioners of Irish Education Reports: Background and Context', in FitzGerald, Garret, *Irish Primary Education in the Early Nineteenth Century* (Royal Irish Academy, 2013)

Ó Cadhain, Máirtín, *Cré na Cille* ('The Dirty Dust' [1949] trans. Alan Titley, Yale University Press, 2015)

O'Callaghan, John, 'Politics, Policy and History: Teaching History in Irish Secondary Schools 1922–1970', *Études irlandaises*, vol. 30, no. 1 (2011), 25–41

O'Casey, Seán, *Juno and the Paycock* (Macmillan and Co., 1925)

O'Connor, Frank, *Irish Miles* (Macmillan, 1947)

_____, *Leinster, Munster and Connaught* (Hale, 1950)

O'Dwyer, Rory, *The Bastille of Ireland: Kilmainham Gaol: From Ruin to Restoration* (The History Press, 2010)

Ó Gráda, Cormac, 'Famine, Trauma and Memory', *Béaloideas*, 69 (2001), 121–43

____, *Eating People is Wrong and Other Essays* (Princeton University Press, 2015)

O'Sullivan, Niamh, *Every Dark Hour: A History of Kilmainham Gaol* (Liberties Press, 2007)

O'Sullivan, Niamh (ed.), *Coming Home: Art and the Great Hunger* (Quinnipiac University Press, 2017)

Oxley, Deborah, *Convict Maids: The Forced Migration of Women to Australia* (Cambridge University Press, 1996)

Pine, Emilie, *The Politics of Irish Memory: Performing Remembrance in Contemporary Irish Culture* (Palgrave Macmillan, 2011)

Pritchett, V. S., *Dublin: A Portrait* (Harper and Row, 1967)

____, *Midnight Oil* (Chatto and Windus, 1971)

Quinn, James, *Young Ireland and the Writing of Irish History* (University College Dublin Press, 2015)

Rae, E. C., 'The Rice Monument in Waterford Cathedral', *PRIA: Archaeology, Culture, History, Literature*, 69 (1970), 1–14

Reilly, Ciarán, *Strokestown and the Great Irish Famine* (Four Courts Press, 2014)

Robinson, Tim, *Connemara: The Last Pool of Darkness* (Penguin, 2009)

Rodgers, Nini, *Ireland, Slavery and Anti-Slavery: 1612–1865* (Palgrave Macmillan, 2007)

Roe, Helen M., 'Cadaver Effigial Monuments in Ireland', *Journal of the Royal Society of Antiquaries of Ireland*, vol. 99, no. 1 (1969), 1–19

Rooney, Brendan (ed.), *Creating History: Stories of Ireland in Art* (Irish Academic Press, 2016)

Rouse, Paul, *Sport in Ireland: A History* (Oxford University Press, 2015)

Ryan, Salvador (ed.), *Death and the Irish: A Miscellany* (Wordwell, 2016)

Sayers, Peig, *Peig: The Autobiography of Peig Sayers*, trans. Bryan McMahon (Talbot Press, 1983)

Shiels, Damian, *The Irish in the American Civil War* (History Press, 2013)

Stone, Philip, 'A Dark Tourism Spectrum: Towards a Typology of Death and Macabre Related Tourist Sites, Attractions and Exhibitions', *Tourism: An Interdisciplinary International Journal*, vol. 54, no. 2 (2006), 145–60

Stone, Philip (ed.), *The Palgrave Handbook of Dark Tourism Studies* (Palgrave, 2018)

Stout, Geraldine, *Newgrange and the Bend of the Boyne* (Cork University Press, 2002)

Tóibín, Colm, *Mad, Bad, Dangerous to Know: The Fathers of Wilde, Yeats and Joyce* (Penguin, 2018)

Tone, Theobald Wolfe, *The Autobiography of Theobald Wolfe Tone*, Seán O'Faolain (ed.) (Thomas Nelson and Sons, 1937)

____, *Life of Theobald Wolfe Tone*, Thomas Bartlett (ed.) (Lilliput Press, 1998)

Toolis, Kevin, *My Father's Wake: How the Irish Teach Us to Live, Love and Die* (Weidenfeld and Nicolson, 2017)

Wadey, Maggie, *The English Daughter* (Sandstone Press, 2016)

Whelan, Kevin, *Religion, Landscape and Settlement in Ireland: From Patrick to Present* (Four Courts Press, 2018)

Whelan, Yvonne, *Reinventing Modern Dublin: Streetscape, Iconography and the Politics of Identity* (University College Dublin Press, 2003)

Woods, C. J., *Bodenstown Revisited: The Grave of Theobald Wolfe Tone, Its Monuments and Pilgrimages* (Four Courts Press, 2018)

Zimmerman, George Dennis, *Songs of Irish Rebellion: Political Street Ballads and Rebel Songs, 1780–1900* (Dublin, 1967, revised edn, Four Courts Press, 2002)

ACKNOWLEDGEMENTS

When I was three my parents, John and Annette, drove from Kerry to Dublin. I spent the journey sticking my head between the driver and passenger seats asking a relentless set of questions. No matter what the answer was, my response was 'But why? But why?' I'm very grateful to my parents for not leaving me on the side of the road that day and for continuing to encourage my incessant inquisitiveness. I have spent my life asking 'But why?' Many years later, in a student flat, I sat with my friends Jude and Regina and we planned to write *Hanging Around Ireland*, a book about trips to former sites of execution. We haven't written it yet, but the intervening years have been full of many late-night conversations about death, misery and the Irish (much less bleak than it sounds). *The Darkness Echoing* owes much both to my parents and to my friendship with Jude and Regina.

I have spent the last two years actively researching, writing and talking about Ireland and dark tourism. For the opportunity to air some of my ideas as I developed them I'm grateful to Gina O'Kelly and the Irish Museums Association, Tamlyn McHugh and everyone involved with the Friends of Sligo Gaol, Craig Slattery and Midlands Science Festival and Jim Carroll of

Acknowledgements

RTÉ Brainstorm and Banter. I am also grateful to Ryan Tubridy, the late Marian Finucane and Neil Delamere for having me on their radio shows to talk about my travels, my grandmother and dark tourism. Special thanks are due to the listeners of *The Ryan Tubridy Show* who contacted me with suggestions of places to visit and people to talk to and to all those who got in touch via Twitter and Instagram. This book is better for their input.

Thanks to my agent Bill Hamilton at A. M. Heath for his advice and support. Thanks to Fiona Murphy of Doubleday Ireland for spotting the potential of this book and for being hugely encouraging throughout. I'm also grateful to Orla King, Sorcha Judge, Jamie O'Connell, Phil Lord, Katrina Whone, Phil Evans, Vivien Thompson, Josh Benn and Marianne Issa El-Khoury at Doubleday Ireland for all their help. And thank you to my copy-editor Sarah Day and to indexer Janet Shuter.

Generally, my research sees me sitting alone in archives and libraries, but researching *The Darkness Echoing* took me to every county in Ireland and over two hundred museums, heritage sites, monuments, memorials, castles, forts, churches, cemeteries and battle sites. I am grateful to everyone I met at sites along the way. I would like in particular to thank those who showed me around their museums and heritage sites, especially Charles Duggan of Dublin City Council, who allowed me a sneak preview of 14 Henrietta Street before it officially opened; Steve Dolan and Mary Healy at the Irish Workhouse Centre; Marie McMahon and Julia Walsh at Tipperary County Museum; Deirdre Bourke at the Independence Museum, Kilmurry; Laura McGreevy at the National Famine Museum, Strokestown Park; Rosemarie

Acknowledgements

Geraghty at Ionad Deirbhile; Ann-Marie Smith at Seamus Heaney: Listen Now Again; Tom O'Neill and John Crotty at Spike Island; Niall Bergin at Kilmainham Gaol; Seán Reynolds and Margaret Ryan at Mountjoy Prison Museum and Jean Wallace at Bunratty Castle and Folk Park. Simon Hill of JANVS | VIDAR encouraged me to take on projects at Kilmainham Gaol and Courthouse and Spike Island, projects which got me thinking about dark tourism in Ireland. Máirtín Ó Catháin and Angela Byrne suggested places to visit in Donegal and Moyra Paterson showed me around Carlow. I am especially grateful to Brenda Malone of the National Museum of Ireland and Brian Crowley of Kilmainham Gaol and Courthouse, who generously shared their knowledge and expertise with me.

Parts of this book were written on my travels, but the bulk of it was completed in Liverpool, Skerries, Dunquin and on Achill Island. For the time and space to plan and write on Achill I am very grateful to the Achill Heinrich Böll Association (and John McHugh and John Smith in particular) for granting me a residency at the Heinrich Böll Cottage. Some of my research trips were funded by Liverpool John Moores University and I am very grateful for that support. I have been lucky to be surrounded by encouraging and supportive colleagues who have made suggestions, read sections and talked nonsense in the Fly in the Loaf, notably Tom Beaumont, David Clampin, James Crossland, Lucy Day, Alison Francis, Susan Grant, Matt Hill, Steve Lawler, Frank McDonough, Alex Miles, Olivia Saunders and Chris Vaughan.

I have been extremely fortunate to have wonderful friends, many of whom have travelled with me to sites, shared stories, read

Acknowledgements

portions of the book, and answered my endless 'What do you know about . . .' questions. There have been many cups of tea drunk, dinners shared and glasses of wine and whiskey consumed, and I am grateful to Jerry Castle, Eric Chenebier, Cristin Colbert, Marie Coleman, Martha Coyne, Michelle Farrell, Sean Farrell, Iseult Fitzgerald, Reachbha FitzGerald, Regina Fitzpatrick, Lisa Griffith, Kate Haldane, Tracey Holsgrove, Agatha Hurst, Eva Kavanagh, David Kirk, Julie Lewis Stauter, Dave Martin, Peter Martin, Jude McCarthy, Stephen McMahon, Jan McVerry, Monnine McCormack, Danielle O'Donovan, Lucy O'Dwyer, Fiacre O'Toole, Jack Paterson, Lorcan Sirr, Michael Staunton, Diane Urquhart and Katie Wink. I am indebted to Anthony Tierney for a very timely intervention and to Barbara Smith who generously allowed me to tell the story of her uncle Thomas Walker.

I am very grateful to Jessie Castle not only for her friendship, advice and for reading the entire manuscript, but also for allowing me to borrow her children and take them with me on many site visits. Thanks to Misha, Matilda and their cousin Lilo for agreeing to this madcap idea and particularly thanks to Misha, who spent weeks travelling around Ireland with me and whose wry observations (if not his music choices) brightened many a long day in the car.

For sharing family tales I am indebted to my parents and my aunt Frances Crowe, who along with my cousin Maura Gallagher had the foresight to make some invaluable recordings of Nana. Thanks also to my cousin Jackie Gorman for sharing stories and allowing me to use one of her poems in the book.

One of the great joys of researching this book was being able

Acknowledgements

to take my nephew Jack and nieces Abbey and Lucy to many of the sites with me. They are the finest research assistants I have ever had. I hope the experience has encouraged them to always ask 'But why?' I look forward to them training up my niece Molly and nephew Billy in time for the next book.

Finally, thanks to my husband, Alistair Daniel, for travelling the length and breadth of Ireland with me (multiple times), for his patience, love and kindness, but most of all for making me laugh every day.

INDEX

Index

Index

Index

Clock Gate Tower, Youghal 219
Clonakilty 52, 87–91
Clonca Church 212–13
Clonmel, siege of 33–4
Clonca Church 212–13
Clonycavan Man 283
Cloonahee Petition 148
Cobh (formerly Queenstown) 121
  Aud scuttled 128
  emigration from 181
  Lusitania survivors & dead taken to 124–6
Cobh Heritage Centre 124–7, 181–4
  Titanic exhibition 121
Coddington family, Oldbridge
  House 40
Coffey, John 121
'coffin ships', emigrant 190
Colley, Edward 184
Collins, Captain Lester 254
Collins, Eileen, (astronaut) 190
Collins, Michael 25
  and Wolfe Tone, considered together 85,
    86, 95–7
  assassinated 86, 87, 90, 92–3
  Crowley on 91–2
  Dublin burial 94–5
    grave 295, 297
  Provisional Government Chair &
    C-in-C 85–6, 97
  relics & memorials 87–96
    Clonakilty statue 88–9
    death mask 96
    memorial, Béal na Bláth 93–4
    Michael Collins film 89, 95
    Michael Collins Centre 91–3
    Michael Collins House, Clonakilty
      89–91
Collins, Seán 92, 94
Collins Barracks Cork Military
  Museum 95
Collins Barracks (Royal Barracks), Dublin
  64, 309
  Collins relics 87
  see also National Museum of Ireland
commemoration kitsch 314–18
Commins, James 148
communities, marginalized, and museums
  329–30
Confederation of Kilkenny 30
Connaught Telegraph 152
Connolly, James 75
  execution 79

Proclamation of republic 73–4
  relics 87
Conolly, William 304
contemporary collecting 331
convicts *see* prisoners; transportation
Coogan, Tim Pat, *The Famine
  Plot* 143
Coosan, 'the Tans' attack 44
Cork City
  Ambassador Hotel 251
  Collins Barracks Cork Military
    Museum 95
  Crawford Gallery 84
  Nano Nagle Place 194–6, 308
  Public Museum 330
  St Fin Barres Cathedral 42–3
  street signs 149
  Triskel Centre Ronan tomb 299
  Williamites besiege 42–3
  see also Elizabeth Fort
Cork City Gaol 158, 222
  ghost convention 302
  prisoners work 156–7, 235
  wedding venue 252
Cork County
  and War of Independence 82–4
  Abbeystrewry Cemetery 150
  Camden Fort Meagher 222
  Charles Fort 302–4
  Clock Gate Tower 219
  Cobh Heritage Centre 121, 124–7,
    181–4
  Cobh Museum 124–6
  Gaol 240–1
  Glanmire 184
  Independence Museum 82–4
  IRA units, anti-treaty 88
  Irish republicanism tradition
    86–91
  Kindred Spirit 165
  Michael Collins Centre 91–3
  Michael Collins House 89–91
  Old Signal Tower 124, 126–8
  Skibbereen Heritage Centre
    150–3
  Titanic Experience 121–4,
    184, 319
  White Star Ticket Office 121
  see also Spike Island
Cork Harbour, British Navy base
  WWI 101–2
'Corkness' 86–7

Index

Index

Index

Index

Index

Index

Index

Index

Index

Index

Index

Index

Index

Index

Dr Gillian O'Brien is Reader in Modern Irish History at Liver-pool John Moores University. She is the author of *Blood Runs Green: The Murder That Transfixed Gilded Age Chicago* and a former Fulbright Scholar.